TWENTY FIRST CENTURY
SCIENCE

Biology for GCSE Combined Science

Project Director
Mary Whitehouse

Project Manager
Alistair Moore

Editor
Alistair Moore

Authors
Neil Ingram
Alistair Moore
Gary Skinner
Mark Winterbottom

Contents

How to use this book

These pages show you all the different features you will find in your Student Book. Each feature is designed to support and develop the skills you will need for your examinations, as well as foster and stimulate your interest in biology.

Chapter openers

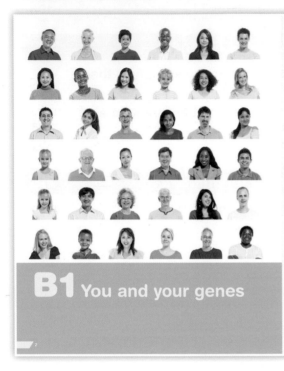

B1 You and your genes

Why study genes?
What makes me the way that I am? How were my features passed on from my parents? What will my children be like? Is my family at risk of genetic diseases? Understanding genes and genomes helps us to answer questions like these. It also helps us develop new technologies that could benefit present and future generations.

What you already know
- Living organisms produce offspring of the same kind, but normally offspring vary and are not identical to their parents.
- Animals reproduce by sexual reproduction. This involves special cells called gametes and the process of fertilisation. Some animals can also reproduce by asexual reproduction.
- Plants reproduce by sexual reproduction. This can involve special structures in flowers, pollination, and fertilisation. Plants can also reproduce by asexual reproduction.
- Genetic information is passed from one generation to the next. This process is called heredity.
- Genetic information is stored in the cell nucleus in structures called chromosomes.
- Chromosomes are made of DNA, and genes are sections of this DNA.
- James Watson, Francis Crick, Maurice Wilkins, and Rosalind Franklin developed a model of the structure of DNA.

The Science
All the information needed to make you is stored in your genetic material. All of your genetic material together is known as your genome. Your genome includes many genes. Almost all of your features depend on your genome and can be affected by your environment. You inherited your genetic information from your parents, and your offspring will inherit genetic information from you.
New gene technologies are allowing us to understand and change some of the information stored in the genome to benefit mankind.

Ideas about Science
Scientists have developed explanations about DNA and the genome. We can use models to make predictions about inheritance.
As we learn more about the genome, new gene technologies are invented. These technologies help us to improve our lives...

Why study
This explains why what you are about to learn is useful to scientists as well as to you.

The Science
This summarises the science in the chapter you're about to study.

Ideas about Science
This provides some ideas about how science explanations are developed and the impact of science and technology on society, in the context of the science in the chapter.

What you already know
This is a summary of the things you've already learnt that will come up again in the chapter. Check through them in advance and see if there is anything that you need to recap.

Main spreads

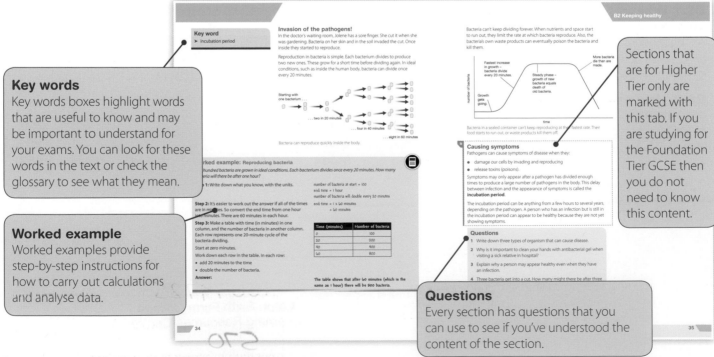

B2 Keeping healthy

Key word
> incubation period

Invasion of the pathogens!
In the doctor's waiting room, Jolene has a sore finger. She cut it when she was gardening. Bacteria on her skin and in the soil invaded the cut. Once inside they started to reproduce.

Reproduction in bacteria is simple. Each bacterium divides to produce two new ones. These grow for a short time before dividing again. In ideal conditions, such as inside the human body, bacteria can divide once every 20 minutes.

Starting with one bacterium...
...two in 20 minutes
...four in 40 minutes
...eight in 60 minutes.

Bacteria can reproduce quickly inside the body.

Worked example: Reproducing bacteria
...hundred bacteria are grown in ideal conditions. Each bacterium divides once every 20 minutes. How many ...teria will there be after one hour?

Step 1: Write down what you know, with the units.
number of bacteria at start = 100
end time = 1 hour
number of bacteria will double every 20 minutes

Step 2: It's easier to work out the answer if all of the times are in minutes. So convert the end time from one hour to minutes. There are 60 minutes in each hour.
end time = 1 x 60 minutes
= 60 minutes

Step 3: Make a table with time (in minutes) in one column, and the number of bacteria in another column. Each row represents one 20-minute cycle of the bacteria dividing.
Start at zero minutes.
Work down each row in the table. In each row:
- add 20 minutes to the time
- double the number of bacteria.

Time (minutes)	Number of bacteria
0	100
20	200
40	400
60	800

Answer:

The table shows that after 60 minutes (which is the same as 1 hour) there will be 800 bacteria.

Bacteria can't keep dividing forever. When nutrients and space start to run out, they limit the rate at which bacteria reproduce. Also, the bacteria's own waste products can eventually poison the bacteria and kill them.

Fastest increase in growth – bacteria divide every 20 minutes.
Steady phase – growth of new bacteria equals death of old bacteria.
Growth gets going.
More bacteria die than are made.

number of bacteria / time

Bacteria in a sealed container can't keep reproducing at the fastest rate. Their food starts to run out, or waste products kill them off.

Causing symptoms
Pathogens can cause symptoms of disease when they:
- damage our cells by invading and reproducing
- release toxins (poisons).

Symptoms may only appear after a pathogen has divided enough times to produce a large number of pathogens in the body. This delay between infection and the appearance of symptoms is called the **incubation period**.

The incubation period can be anything from a few hours to several years, depending on the pathogen. A person who has an infection but is still in the incubation period can appear to be healthy because they are not yet showing symptoms.

Questions
1 Write down three types of organism that can cause disease.
2 Why is it important to clean your hands with antibacterial gel when visiting a sick relative in hospital?
3 Explain why a person may appear healthy even when they have an infection.
4 Three bacteria get into a cut. How many might there be after three...

Key words
Key words boxes highlight words that are useful to know and may be important to understand for your exams. You can look for these words in the text or check the glossary to see what they mean.

Worked example
Worked examples provide step-by-step instructions for how to carry out calculations and analyse data.

Sections that are for Higher Tier only are marked with this tab. If you are studying for the Foundation Tier GCSE then you do not need to know this content.

Questions
Every section has questions that you can use to see if you've understood the content of the section.

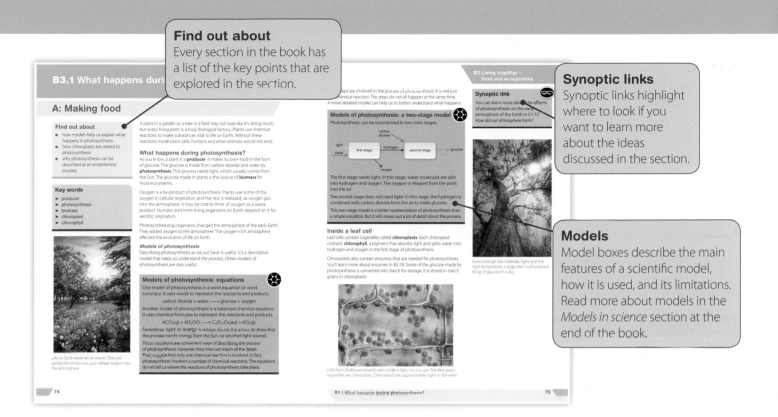

Find out about
Every section in the book has a list of the key points that are explored in the section.

Synoptic links
Synoptic links highlight where to look if you want to learn more about the ideas discussed in the section.

Models
Model boxes describe the main features of a scientific model, how it is used, and its limitations. Read more about models in the *Models in science* section at the end of the book.

Science explanations and Ideas about Science

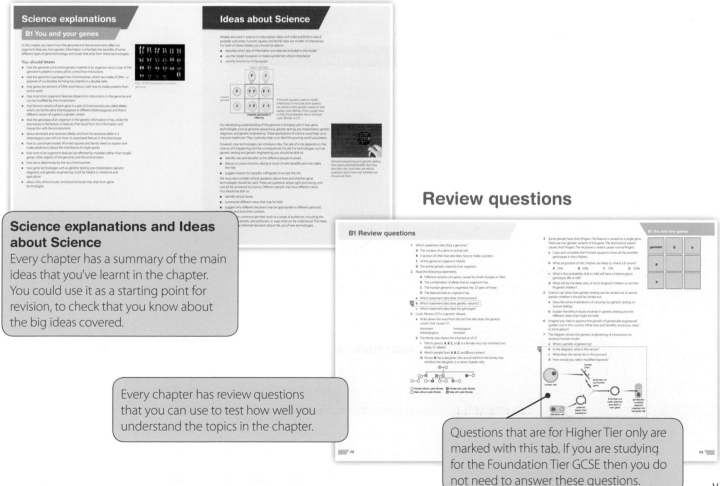

Science explanations and Ideas about Science
Every chapter has a summary of the main ideas that you've learnt in the chapter. You could use it as a starting point for revision, to check that you know about the big ideas covered.

Review questions

Every chapter has review questions that you can use to test how well you understand the topics in the chapter.

Questions that are for Higher Tier only are marked with this tab. If you are studying for the Foundation Tier GCSE then you do not need to answer these questions.

Structure of assessment

There will be four examination papers for GCSE Combined Science.

All of the papers cover all of the chapters. They contain short-answer questions worth up to three marks, structured questions, and questions requiring an extended written response. The papers will assess theory, problem solving, calculations, and questions about practical work.

Paper 1: Biology

Time	Marks available	Percentage of GCSE
1 hour 45 minutes	95	26.4%

Paper 2: Chemistry

Time	Marks available	Percentage of GCSE
1 hour 45 minutes	95	26.4%

Paper 3: Physics

Time	Marks available	Percentage of GCSE
1 hour 45 minutes	95	26.4%

Paper 4: Combined Science

Time	Marks available	Percentage of GCSE
1 hour 45 minutes	75	20.8%

Top tips

In each examination paper, you will have to:

- demonstrate your knowledge and understanding of scientific ideas, techniques, and procedures

- apply your knowledge and understanding to new and familiar contexts

- interpret and evaluate information and data, make judgements and conclusions, and evaluate scientific procedures and suggest how to improve them.

Some questions will use contexts that you are not familiar with, including examples of science from the real world, issues from the news, and reports of scientific investigations. Remember that although the context may be different, the science is the same. The questions are designed so that you can answer them if you combine your own knowledge and understanding with the information given in the question. You should:

- think about how the context is similar to something you have learnt about

- look for information in the question that suggests how you can relate what you know to the new context.

Kerboodle

This book is also supported by Kerboodle, offering unrivalled digital support for building your practical, maths, and literacy skills.

If your school subscribes to Kerboodle, you will find a wealth of additional resources to help you with your studies and with revision:

- Animations, videos, and podcasts

- Webquests

- Activities for every assessable learning outcome

- Maths skills activities and worksheets

- Literacy skills activities and worksheets

- Interactive quizzes that give question-by-question feedback

- Ideas about Science case studies

Watch interesting animations on the trickiest topics, and answer questions afterwards to check your understanding.

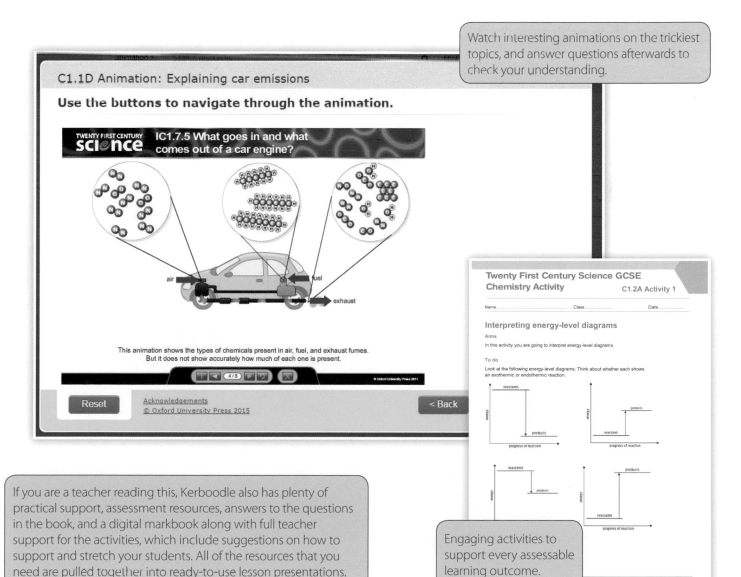

If you are a teacher reading this, Kerboodle also has plenty of practical support, assessment resources, answers to the questions in the book, and a digital markbook along with full teacher support for the activities, which include suggestions on how to support and stretch your students. All of the resources that you need are pulled together into ready-to-use lesson presentations.

Engaging activities to support every assessable learning outcome.

B1 You and your genes

Why study genes?

What makes me the way that I am? How were my features passed on from my parents? What will my children be like? Is my family at risk of genetic diseases? Understanding genes and genomes helps us to answer questions like these. It also helps us develop new technologies that could benefit present and future generations.

What you already know

- Living organisms produce offspring of the same kind, but normally offspring vary and are not identical to their parents.
- Animals reproduce by sexual reproduction. This involves special cells called gametes and the process of fertilisation. Some animals can also reproduce by asexual reproduction.
- Plants reproduce by sexual reproduction. This can involve special structures in flowers, pollination, and fertilisation. Plants can also reproduce by asexual reproduction.
- Genetic information is passed from one generation to the next. This process is called heredity.
- Genetic information is stored in the cell nucleus in structures called chromosomes.
- Chromosomes are made of DNA, and genes are sections of this DNA.
- James Watson, Francis Crick, Maurice Wilkins, and Rosalind Franklin developed a model of the structure of DNA.

The Science

All the information needed to make you is stored in your genetic material. All of your genetic material together is known as your genome. Your genome includes many genes. Almost all of your features depend on your genome and can be affected by your environment. You inherited your genetic information from your parents, and your offspring will inherit genetic information from you.

New gene technologies are allowing us to understand and change some of the information stored in the genome, to benefit mankind.

Ideas about Science

Scientists have developed explanations about DNA and the genome. We can use models to make predictions about inheritance.

As we learn more about the human genome, new gene technologies are invented. These technologies could help us to improve healthcare and ensure we can feed the growing human population. However, we must decide how and whether these technologies should be used. This involves decisions about what is right and wrong, and science cannot provide all the answers.

A: What is your genome?

Find out about

- genetic material and the genome
- chromosomes
- the structure of DNA
- how your genome and your environment affect your features

Key words

- ➤ inherited
- ➤ genome
- ➤ nucleus
- ➤ eukaryotic organism
- ➤ prokaryotic organism
- ➤ plasmid
- ➤ chromosome
- ➤ DNA

A lot of information goes into making a human being. All people have most features in common. The differences between us are very small – but they are what make us unique.

The same and different

Georgia and Paige are sisters. They look similar to each other and their parents, but there are differences between them.

Humans, other animals, and plants look a lot like their parents. This is because they have **inherited** genetic information from them. This information controls how an organism develops and functions. It is stored in genetic material. All the genetic material of an organism is called its **genome**.

Sisters Georgia and Paige inherited their genomes from their parents. Each girl inherited a slightly different combination of genetic material from their mother and their father. This is one reason why there are differences between them.

All of the information needed to make each girl is contained in her genome. Most body cells contain a complete copy of the genome.

Genomes and the environment

You may have features that are the result of only your environment. One example is physical damage, such as a scar.

But almost all of your features are affected by both the genetic information you inherited and your environment. For example, your body mass depends on information in your genome but you will become heavier if you eat too much. This is why it's important to eat a healthy diet and exercise regularly.

Your environment does not interact directly with your genome. The environment outside a cell can affect molecules on the cell surface. This can start a chain reaction of changes in the cell cytoplasm. Eventually, this can affect how the information in the genome is used – and this can affect your features.

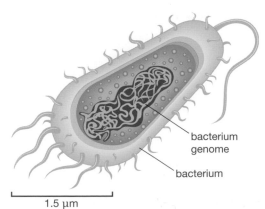

bacterium genome

bacterium

1.5 μm

All living organisms have a genome, from bacteria to humans.

The features of each student are influenced by the genetic information they inherited and by their environment.

Genomes in different organisms

Genetic material in humans and other animals is found inside the **nucleus** of each cell. This is also true for plants, fungi, and protists. Each of these organisms is called a **eukaryotic organism** because its cells have a nucleus.

Bacteria are simpler organisms. Their cells do not have a nucleus and the genetic material is located in the cytoplasm. A bacterium is called a **prokaryotic organism** because it does not have a nucleus. Some bacteria contain small loops of genetic material. Each loop is called a **plasmid**. Plasmids can be used in genetic engineering – you will learn more about this in B1.3.

Inside the nucleus

If you look at stained animal or plant cells using a light microscope you can see the nucleus of each cell as a dark blob. A light microscope is not powerful enough to see the nucleus in any more detail. If you could see inside the nucleus, you would see the genetic material. It is stored as long threads called **chromosomes**.

There are 46 chromosomes in the nucleus of a human body cell. Chromosomes are present as pairs. In a human cell there are 23 pairs. Chromosomes are made of very long molecules of **DNA**.

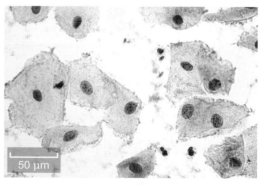

These human cheek cells have been stained with a solution of dye and then examined using a light microscope. The nucleus of each cheek cell is darkly stained.

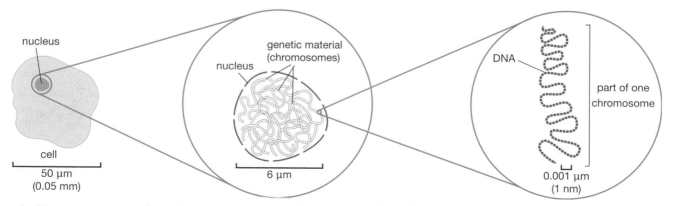

All of the genetic material needed to create a human fits into the nucleus of a cell.

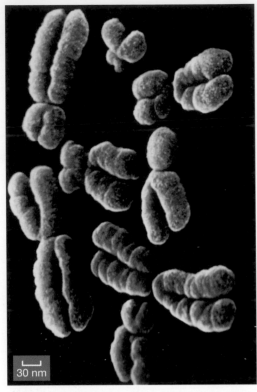

An image of human chromosomes taken using an electron microscope. Each chromosome is tightly coiled up like this when a cell gets ready to divide.

James Watson (left) and Francis Crick with a model of the structure of DNA.

Key words

➤ double helix
➤ nucleotide

What is DNA?

The structure of DNA was discovered in 1953 by James Watson and Francis Crick. They interpreted a photograph of DNA that had been taken by Rosalind Franklin and Maurice Wilkins using X-rays. They saw a repeating pattern that helped them to explain the structure.

DNA consists of two strands that are twisted together to form a **double helix**. Each strand of DNA is a long molecule made from lots of small molecules joined together in a chain. The small molecules are called **nucleotides**.

DNA is an example of a natural polymer. A polymer is a long chain molecule made from lots of small molecules, called monomers, joined together. Nucleotides are monomers that join together to make the DNA polymer.

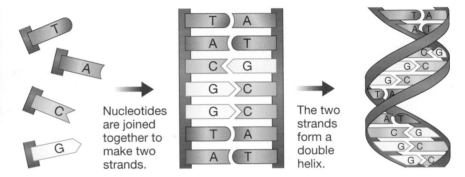

DNA is a polymer made up of nucleotides, which form two strands in a double helix.

Questions

1 Choose two students in the photograph on page 5.
   a Write down five ways in which the students look different from one another.
   b What could have caused the differences between the two students?

2 Where would you find your genome?

3 Explain the difference between a eukaryotic cell and a prokaryotic cell.

4 Copy and complete the following sentences.
   a The cell nucleus contains long threads of genetic material called . . .
   b The threads are made of very long molecules of . . .

5 Explain how your genome can be affected by the environment.

B: Genes and proteins

The information stored in Georgia's genome controls most of her features. But how does it do this? The answer lies in **genes** and proteins.

What are genes?

A gene is a region of DNA. Each of your chromosomes contains many genes. Genes contain the instructions that cells use to make proteins. Each cell nucleus contains the instructions for making every protein in a human being. But in any one cell nucleus only some genes are being used to make proteins – the rest are inactive.

We think there are about 20 000 genes in the human genome. Genes are very important, but they only make up about 1.5% of your genome. The remaining 98.5% of your DNA is more mysterious, and for a time scientists described it as 'junk'. Now scientists think that up to 80% of this DNA is important in controlling **gene expression**. This means it controls when the information in genes is used to make proteins. The study of the structure and function of genomes is called **genomics**. This is an exciting and fast-moving area of science.

We do know that the other parts of the genome are important in controlling when and how genes are used to make proteins.

What are proteins?

Proteins are important molecules. There are many different proteins, which can be:

- structural – for example collagen (a protein found in connective tissues)
- functional – for example amylase (an enzyme that speeds up chemical reactions).

Protein molecules are polymers made from **amino acids**. Amino acid monomers are joined together by chemical bonds. They can be arranged in different sequences to make all of the proteins in our bodies. Small proteins might contain about 50 amino acids but some complex proteins contain thousands.

The amino acid glycine is made from carbon (**C**), hydrogen (**H**), oxygen (**O**), and nitrogen (**N**). Other amino acids have more complicated structures, but they all have these elements in common.

Different proteins have different shapes. The exact shape of a protein can be very important to how it works. The shape of a protein molecule is controlled by the instructions stored in genes.

The effects of genes on development

The information stored in many genes and other regions of your genome is used to control your growth.

But a single gene can have a major disruptive effect on development. For example, a change in a gene called *PAH* in the human genome can lead to a disease called phenylketonuria (PKU). We can learn a lot about how genes work from this example.

Find out about

- genes, alleles, and genetic variants
- genotype and phenotype
- making proteins

Key words

- ➤ gene
- ➤ gene expression
- ➤ genomics
- ➤ amino acid

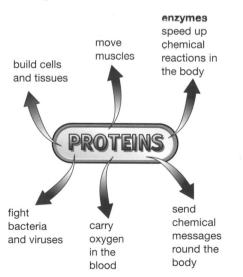

Georgia's features depend on the information in her genome.

enzymes speed up chemical reactions in the body

move muscles

build cells and tissues

fight bacteria and viruses

carry oxygen in the blood

send chemical messages round the body

There are about 50 000 different proteins in the human body.

What is phenylketonuria?

A person who has the disease PKU cannot make an important enzyme. The enzyme is usually made in the liver to break down a substance called phenylalanine. This substance is found in food and is sometimes used to artificially sweeten drinks. When the enzyme does not work properly, phenylalanine can build up in the blood and brain. In young children and babies this can harm the development of the brain and nervous system.

Why is the PAH gene important?

There are about a thousand genes on chromosome 12. One of these genes contains the information to make the enzyme that breaks down phenylalanine. This is the *PAH* gene.

You have two copies of chromosome 12 in your genome. You inherited one copy from your mother and the other from your father. Both copies carry the *PAH* gene. The two copies of a gene are called **alleles**, and they may contain exactly the same or different information.

A different version of a gene is a **genetic variant**. There are about 500 known variants of the *PAH* gene. These variants are caused by small changes in the DNA. Most of these changes do not affect the way the enzyme works, but some of the variants are harmful – they change the shape of the enzyme, which stops it from working.

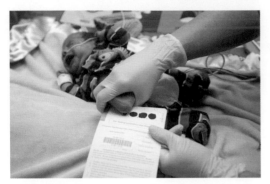

The blood of newborn babies is always tested for PKU using the heel-prick test. Babies with the disease are put on a diet free of phenylalanine so that their brains will develop normally.

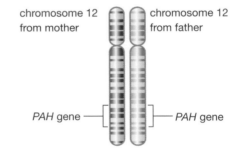

If both alleles in a pair of chromosomes are harmful variants of the *PAH* gene, the person will have the symptoms of PKU.

Genotype and phenotype

The genetic variants that an organism has make up its **genotype**. The genotype for a single gene is the combination of alleles for that gene. The features that result from the information in the DNA (and its interaction with the environment) make up an organism's **phenotype**.

In the *PAH* example:

● the two *PAH* alleles are the genotype

● a person with two harmful *PAH* alleles has the PKU phenotype (they have the disease).

Putting the story together: from genotype to phenotype

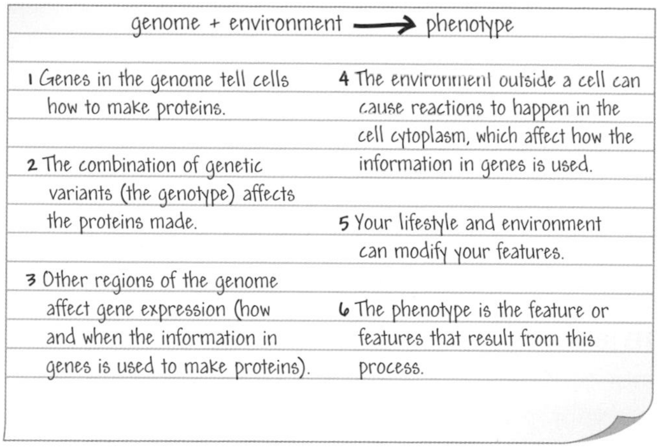

genome + environment ⟶ phenotype	
1 Genes in the genome tell cells how to make proteins.	4 The environment outside a cell can cause reactions to happen in the cell cytoplasm, which affect how the information in genes is used.
2 The combination of genetic variants (the genotype) affects the proteins made.	5 Your lifestyle and environment can modify your features.
3 Other regions of the genome affect gene expression (how and when the information in genes is used to make proteins).	6 The phenotype is the feature or features that result from this process.

Notes about how an organism's genome and environment affect its features.

Questions

1 Write down two jobs done by proteins in the human body.

2 Write these cell parts in order from smallest to largest: chromosome, gene, nucleotide, nucleus.

3 What is the difference between your genotype and your phenotype?

4 Describe the relationship between proteins and amino acids.

5 Explain how an organism's genes and environment can affect its features.

Key words

➤ allele
➤ genetic variant
➤ genotype
➤ phenotype

A: Explaining inheritance

Find out about

- gametes
- dominant and recessive alleles
- homozygous and heterozygous genotypes
- predicting how single genes will be inherited

Key words

➤ gamete
➤ dominant
➤ recessive
➤ homozygous
➤ heterozygous

This baby has inherited a unique mix of genetic information.

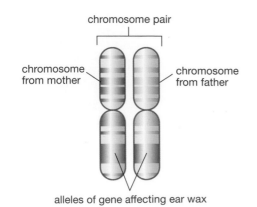

A pair of chromosomes. The gene affecting ear wax is coloured pink. The two alleles of the gene are in the same place on each chromosome.

You may have heard the saying 'it runs in the family'. People in a family usually look like one another. Perhaps you don't like a feature you've inherited – your dad's big ears or your mum's short fingers. Have you ever wondered how you inherited the information that gave you your features?

The role of gametes

The information in your genome was copied from your parents. You inherited half of it from your mother and half from your father. If you become a parent, each of your children will inherit half of their genome from you.

This genetic information is passed on by **gametes** – an egg cell from the mother and a sperm cell from the father. Gametes are made from body cells. They have half the number of chromosomes as body cells. When a sperm fertilises an egg, the resulting cell has the correct number of chromosomes for a body cell.

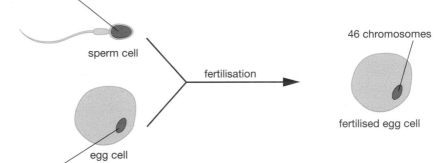

Gametes (sperm and egg cells) have half the number of chromosomes as body cells. The fertilised egg cell has the same number of chromosomes as body cells. This diagram is not to scale – a human egg cell is 20-times larger than a human sperm cell.

A few features are controlled by a single gene

A very small number of your features are controlled by a single gene. These can help us understand how genes are inherited.

Ear wax

Ear wax may not be something you want to think much about. But the type of ear wax you have is a good example of a feature controlled by a single gene.

The gene is called *ABCC11*. One genetic variant of this gene causes wet ear wax. Another variant causes dry ear wax. Which type of ear wax you make depends on which genetic variants you inherited. The difference between the two variants is a point mutation. One adenine base (in the dry variant) is changed to guanine (in the wet variant).

Dominant alleles – they're in charge

We can represent the two variants of the ear wax gene using letters. For example:

● The E variant causes wet ear wax.

● The e variant causes dry ear wax.

The E variant is **dominant**. You only need one dominant variant in a pair of alleles to have its feature (in this case, wet ear wax). The e variant is **recessive**. You must have two copies of a recessive variant in a pair of alleles to have its feature (in this case, dry ear wax).

A person with two alleles that are the same has a **homozygous** genotype (e.g., EE or ee). A person who has two different alleles (e.g., Ee) has a **heterozygous** genotype. The type of ear wax is the phenotype, which is linked to the genotype.

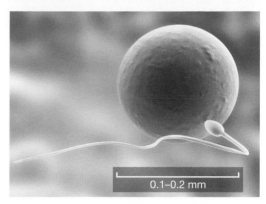

A computer-generated illustration of a human sperm cell about to fertilise a human egg cell. A fertilised egg cell (called a zygote) has genetic information from both parents.

This person inherited a dominant E variant from both parents. They are homozygous. They have wet ear wax.

This person inherited a recessive e variant from both parents. They are homozygous. They have dry ear wax.

This person inherited one E variant and one e variant. They are heterozygous. They have wet ear wax.

Which alleles are passed on by the gametes?

Gametes have a copy of one chromosome from each chromosome pair in the body cells. Imagine a parent who is homozygous for the wet ear wax variant. Their genotype is EE. All the gametes they produce will have the E variant. In a parent who is ee homozygous, all the gametes will have the e variant.

But if a parent is heterozygous (Ee), they will produce two sorts of gamete. Half the gametes will have the E variant. The other half will have the e variant.

Wet ear wax, shown here, is moist and yellow-brown in colour. Dry ear wax is flaky and a much paler colour.

Modelling the inheritance of single genes

Two parents want to have a child. We don't know which egg and which sperm will meet at fertilisation. But we can use a genetic diagram to model all the possible combinations. This kind of diagram is called a **Punnett square**.

Modelling inheritance using the Punnett square

A Punnett square is used to model the inheritance of alleles of a single gene. It is a mathematical model. It uses data about the genotype of the mother and the genotype of the father. It assumes that half of the mother's gametes have one allele from a pair, and the other half have the other allele. It assumes the same thing for the father.

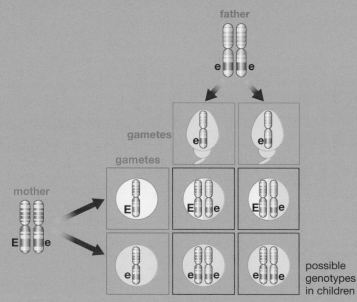

The Punnet square models all of the possible combinations of alleles at fertilisation. We can use this to predict how many of the children are likely to have particular genotypes.

This type of Punnett square only works for the inheritance of single genes, but most of your features depend on more than one gene. More complicated models are needed to model the inheritance of most features.

In the Punnett square in the model box, two out of the four possible genotypes in the children are Ee. The genotype Ee causes the wet ear wax phenotype.

- We can write 2 out of 4 as $\frac{2}{4}$, which is the same as $\frac{1}{2}$, which is equal to 50% and equal to 0.5.
- It is likely that 50% of the children will have the genotype Ee.
- The probability that any child will have the genotype Ee is 0.5.

The same is true for the genotype ee. The genotype ee causes the dry ear wax phenotype.

The ratio of genotypes Ee:ee in the children is likely to be 2:2, which is the same as 1:1. If the couple have six children, it is likely that three will be Ee and three will be ee. Therefore, we could conclude that half their children will have wet ear wax and half will have dry ear wax.

These are just **predictions** based on the probabilities and ratio suggested by the Punnet square model. We cannot be certain how many of the children will have each genotype. We can only be sure of a child's genotype after fertilisation.

The Punnett square is exactly the same each time fertilisation occurs. So for each child, the probability of having a particular genotype is not affected by previous children the parents have had. It is possible, though unlikely, that the parents could have six children all with the genotype Ee.

Another useful, simple model of inheritance is a **family tree**. It is another type of genetic diagram.

Modelling inheritance using the family tree

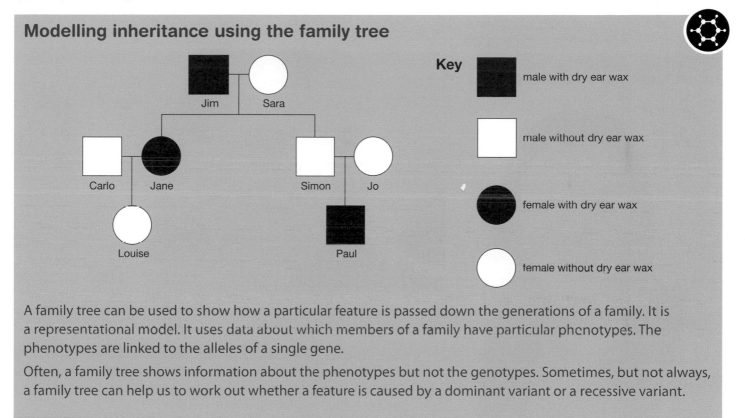

A family tree can be used to show how a particular feature is passed down the generations of a family. It is a representational model. It uses data about which members of a family have particular phenotypes. The phenotypes are linked to the alleles of a single gene.

Often, a family tree shows information about the phenotypes but not the genotypes. Sometimes, but not always, a family tree can help us to work out whether a feature is caused by a dominant variant or a recessive variant.

The family tree in the model box shows which members of a family have dry ear wax. Note that:

● grandfather Jim has dry ear wax

● son Simon does not have dry ear wax

● grandson Paul has dry ear wax.

The feature seems to skip a generation. How can this be possible?

We know that the genetic variant that causes dry ear wax is recessive. A person will only have dry ear wax if they have two copies of the variant e, that is, they have the homozygous genotype ee. If a person has the genotype Ee or EE, they will not have dry ear wax. Paul must have the genotype ee, because he has dry ear wax. So he inherited an e from Simon and an e from Jo. Neither Simon nor Jo has dry ear wax, so they must both have the genotype Ee. They are both **carriers** of the recessive variant.

Key words

➤ Punnett square
➤ prediction
➤ family tree
➤ carrier

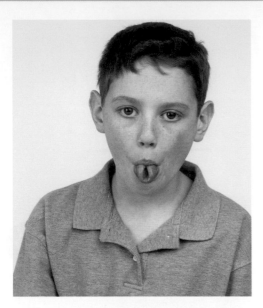

The ability to roll your tongue is another feature that we now think is controlled by multiple genes, not just one.

Carriers:

- have a recessive variant
- can pass the recessive variant on to their children
- do not show the recessive variant's phenotype, because they are heterozygous.

Most features are not controlled by a single gene

Very few of your features are controlled by a single gene. In fact, scientists have changed their thinking about a number of features, such as dimples, curved thumbs, and the ability to roll your tongue. Scientists used to think these features were controlled by a single gene. Now we know the story is more complicated.

This boy has dimples. We used to think this feature was controlled by a single gene. We now think it is controlled by multiple genes.

Most of your features depend on multiple genes and on other regions of the genome. More complicated models are needed to predict how inheritance will affect these features. You may learn more about this if you keep studying biology after your GCSEs.

Questions

1 Explain the difference between a dominant variant and a recessive variant.

2 Look at the family tree in the family tree Modelling box.
   a What is Jane's phenotype?
   b What is Jane's genotype?

3 A mother has the genotype Ee. A father has the genotype EE.
   a Which person, the mother or the father, is heterozygous?
   b Draw a Punnett square to show all the possible genotypes in their children.
   c What is the probability that a child will have wet ear wax?

4 Write down one limitation of a Punnett square as a model of inheritance.

B: Male or female?

For you, it all began at the moment of fertilisation. The nucleus of your father's sperm joined with the nucleus of your mother's egg. Your genome formed and your **sex** – whether you are male or female – was decided. Your sex is determined by biological factors such as genes and hormones.

Your **gender** is more complicated. It depends on whether you feel masculine or feminine, and whether you identify yourself as a man or a woman. Your gender may be influenced by many things, including your own feelings, the culture you live in, your family, and biological factors.

What decides an embryo's sex?

A fertilised human egg cell has 23 pairs of chromosomes. Pair 23 are the **sex chromosomes**. Males have an X chromosome and a Y chromosome in the pair (XY). Females have two X chromosomes (XX).

Chromosomes are present in pairs. When gametes are made they only get one chromosome from each pair. So sperm cells can get an X chromosome or a Y chromosome. All egg cells get an X chromosome.

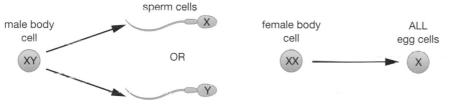

When a sperm cell fertilises an egg cell there is a 50% chance it will be a sperm with an X chromosome. There is an equal chance it will be a sperm with a Y chromosome. This means there is a 50% chance the baby will be a male and a 50% chance it will be a female.

What makes a baby male?

Male and female embryos are very alike until they are about six weeks old. Then a male embryo's testes start to develop. This is caused by a gene on the Y chromosome – the *SRY* gene.

The Y chromosome carries the *SRY* gene that triggers the development of testes.

Testes produce male **sex hormones** called androgens. Androgens make the embryo develop male characteristics. If there is no male sex hormone present, the embryo develops the ovaries, clitoris, and vagina of a female.

Hormones are a group of proteins that control many processes in cells. Tiny amounts of hormones are made by different parts of the body. They travel to their target cells in the bloodstream.

Find out about

- what decides if you are male or female
- the role of sex chromosomes
- the difference between sex and gender

Key words

- ➤ sex
- ➤ gender
- ➤ sex chromosomes
- ➤ sex hormone

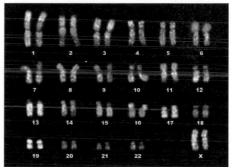

The 23 pairs of chromosomes of a human female. Female humans have two X chromosomes.

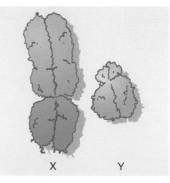

Male humans have an X and a Y chromosome.

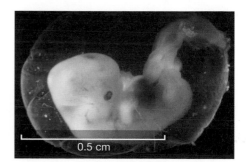

0.5 cm

This embryo is six weeks old.

Jan's story

Jan having fun on holiday.

At the age of 18 Jan was at university. She was very happy, and was going out with a football player. She thought her periods hadn't started because she did a lot of sport. She discussed it with her doctor, who arranged for her to have some tests.

The tests involved Jan giving a sample of her cheek cells for analysis. She discovered that she had male sex chromosomes (XY).

Sometimes a person has X and Y chromosomes but looks female. This is because their body makes androgens but the cells don't respond to them. About 1 in 20 000 people have this condition. It is called androgen insensitivity syndrome. These people have small internal testes and short vaginas. They can't have children.

Jan had no idea she had this condition. She found it difficult to come to terms with it at first. But now she has told her boyfriend and they have stayed together. She is happy with her feminine gender identity.

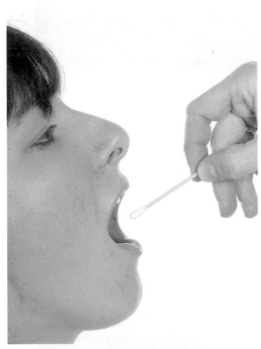

A genetic test can use cheek cells in saliva as a source of DNA.

Questions

1 Which sex chromosome(s) would be in the nucleus of:
   a a male human's body cell
   b an egg cell
   c a female human's body cell
   d a sperm cell?

2 Explain how the Y chromosome makes an embryo male.

B1.3 How can and should gene technology be used?

A: Testing choices

People who are planning to become parents want to have a healthy baby. They try to do all the right things to give their baby a healthy life. But doing the right thing is not always easy. There can be difficult choices to make when genes are involved.

Scientists are unlocking the secrets of the human genome. Comparing the genomes of people with and without a disease can help us find genetic variants linked to the disease. We now know that some diseases are caused by harmful genetic variants. PKU, Huntington's disease, and cystic fibrosis (CF) are all caused by single genes. Huntington's disease is caused by a dominant allele. CF is caused by a recessive allele.

What causes cystic fibrosis?

CF is not caused by an infection. It is an inherited disease caused by a genetic variant of the *CFTR* gene. It is the most common disease in Europe caused by a genetic variant.

The disease causes big problems for breathing and digestion. Cells that make mucus in the body don't work properly. The mucus is much thicker than it should be so blocks up the lungs. It also blocks tubes that take enzymes from the pancreas to the gut. People with CF get breathless. They also get lots of chest infections. The shortage of enzymes in their gut means that their food isn't digested properly so they can be short of nutrients.

Inheriting cystic fibrosis

Most people with CF can't have children. The thick mucus affects their reproductive systems. So babies with CF are usually born from healthy parents. The genetic variant that causes CF is recessive. The parents of people with CF are carriers – they have one healthy variant (F) and one harmful variant (f). When two gametes carrying the faulty allele meet at fertilisation, the baby will have CF.

Genetic testing and cystic fibrosis

About one in every 25 people in the UK carries the harmful f variant that causes CF. About one baby in every 2500 is born with CF. At present there is no cure. CF is a disease that has to be managed for the whole of a person's life.

Can **gene technology** help us make better decisions about inherited diseases? Most parents who give birth to a child with CF have no idea they are carriers of the harmful variant. It is easy to test adults to see whether their genome contains the harmful variant. The test can be done using a sample of saliva. This is an example of **genetic testing**. It is a type of gene technology.

Find out about

- how our increasing understanding of the human genome affects family planning
- genetic testing as an example of gene technology
- risk
- making ethical decisions

Key words

➤ gene technology
➤ genetic testing

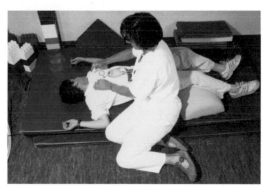

This boy has CF. He has physiotherapy every day to clear thick mucus from his lungs.

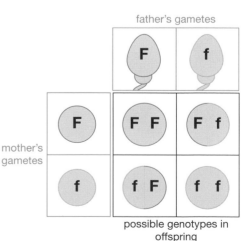

The recessive f variant causes CF. There is a 25% chance that a child from carrier parents will have the disease.

If there is a history of CF in a family, then genetic testing may be a good idea. But the test may just be the beginning of the story. If a couple discover that they are both carriers, should they go ahead and try for a baby?

Elaine and Peter's story

Elaine and Peter wanted to have a baby. But Elaine's nephew has CF. Elaine and Peter were worried that their child may also be born with the disease. They both had a genetic test. The tests showed that they are both carriers of the harmful f allele for CF.

The doctor explained to Elaine and Peter that there was a 25% chance their baby would have CF. This meant there was a 75% chance the baby would not have CF. Elaine and Peter knew there was a **risk** their child could have the disease. The doctor explained that the size of a risk always depends on:

● the chance of it happening

● the consequences if it did happen.

In this case, the chance of having a baby without CF was three times greater than the chance of having a baby with CF. But they also had to think about the consequences of CF for the child, and for themselves. They knew that caring for a child with CF would be a lot of extra responsibility.

Some questions can be answered by science. For example, the question of whether Elaine and Peter were at risk of having a child with CF. But other questions cannot be answered by science. Only Elaine and Peter could decide whether or not to have children. They decided to try for a baby.

Another type of test

When Elaine became pregnant, the doctor told her they could do a genetic test on cells from the fetus. This would show whether or not it had CF. There are two ways of testing fetal cells:

● an amniocentesis test, which tests cells from amniotic fluid

● a chorionic villus test, which tests cells from the placenta.

Elaine and Peter get advice from a genetic counsellor.

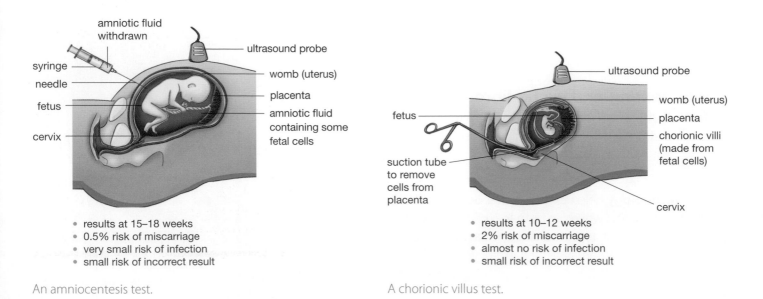

amniotic fluid withdrawn
ultrasound probe
syringe
womb (uterus)
needle
fetus
placenta
cervix
amniotic fluid containing some fetal cells

● results at 15–18 weeks
● 0.5% risk of miscarriage
● very small risk of infection
● small risk of incorrect result

An amniocentesis test.

ultrasound probe
fetus
womb (uterus)
placenta
chorionic villi (made from fetal cells)
suction tube to remove cells from placenta
cervix

● results at 10–12 weeks
● 2% risk of miscarriage
● almost no risk of infection
● small risk of incorrect result

A chorionic villus test.

The tests are applications of science that give us useful information. They help us make informed decisions. But there are many other factors to consider. The tests themselves are not free of risks. Elaine decided to have an amniocentesis test. The results showed that the fetus had CF.

Ethics – right and wrong

Elaine and Peter faced a very difficult decision. Should they continue with the pregnancy? Or should they terminate it (have an abortion)? When a person has to make a decision about what is the right or wrong way to behave, they are thinking about an **ethical** question.

Elaine and Peter had to weigh up the consequences of each choice. They thought about how each choice would affect all the people involved. They had to make judgements about the difficulties they and their child would face if they continued with the pregnancy.

Some people think CF would have an unacceptable effect on a person's quality of life. CF cannot be cured, but treatments are improving and life expectancy is getting better all the time. Many people lead happy, full lives in spite of illnesses. Some people think that barriers created by society, rather than disease, are the cause of disability.

Others think having a termination is wrong and should never be done. They believe an unborn child has the right to life. People may have different views because of their personal beliefs. Not everyone weighing up the consequences of each choice would come to the same decision.

This is the hardest decision that Elaine and Peter ever had to make. They decided to continue with the pregnancy and care for their child.

Do genetic tests always give the correct result?

With every genetic test, there is a small chance that it will give an incorrect result. This is a practical limitation of the test.

- A **false negative** test fails to detect the presence of a harmful genetic variant when it is present.

- A **false positive** test diagnoses a harmful genetic variant when it is not present.

The risk of a false result must be considered when deciding whether to have a genetic test, and what to do with the results.

Some things are wrong and should never be done.

The right decision is the one that leads to the best outcome for the most people.

Is it fair for us to have a baby, knowing it will be ill?

It's wrong to have a termination. We'll look after our baby whatever happens.

It's wrong to assume the child will suffer just because it has a genetic disease.

Key words

➤ risk
➤ ethical
➤ false negative
➤ false positive

What could be done? What should be done?

Elaine and Peter faced a very difficult choice. Now doctors can offer a different option. It is called **pre-implantation genetic diagnosis (PGD)**.

Eggs are removed from a mother's ovaries. They are fertilised with the father's sperm in a laboratory. When the embryos are 3–6 days old, a cell is removed from each embryo. The cells are tested for harmful genetic variants. An embryo without harmful variants is selected and implanted into the mother's uterus. If successful, the child (and later their own children) will be free of the genetic disease.

The technique has risks:

● Collecting the eggs is difficult and unpleasant for the woman.

● The fertilisation and implantation does not always work.

● Removing a cell from an embryo might weaken it.

● There is a chance of a false result from the genetic test.

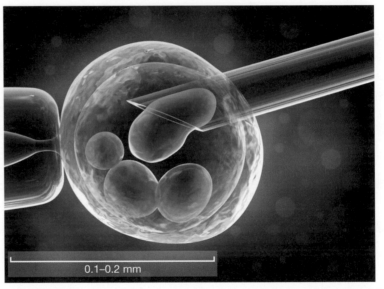

0.1–0.2 mm

In PGD, a cell is removed from an embryo. Its DNA is tested for harmful genetic variants.

Some people argue that it is wrong to select certain embryos and destroy the others. What do you think? Science can suggest many things that could be done. But this does not always mean they should be done. In this world of testing choices, there are no easy options.

Questions

1 Write down the names of two ways to test the genome of a fetus.

2 What are 'false negative' and 'false positive' test results?

3 What must be considered when estimating the size of a risk?

4 Explain what it means when somebody is a carrier of CF.

5 Draw a table summarising arguments for and against PGD.

B: Medicine made personal

Identical twins are born with identical genomes. For the rest of us, about 99.9% of the information in our genome is the same as every other person on the planet. The remaining 0.1% is what makes us unique, but it can make a big difference.

Some of your genetic variants may cause disease. Or they may affect the way you react to certain medicines. Could doctors use information about our genomes to make sure we get the right medicines and treatments?

Nisha's story

In 2015, Nisha was diagnosed with cancer. She started to take a medicine she hoped would help cure her. But four days later she was in intensive care with her heart struggling to keep going. She is one of a small number of people whose bodies react badly to a particular medicine. Nisha recovered with no permanent damage to her heart – and so far the cancer hasn't returned either.

Personalised medicine

In the future genetic testing could make problems like Nisha's a thing of the past. Some people produce enzymes that break down medicines very quickly. They need higher doses of a medicine than most of us. Other people don't produce the enzymes they need to break down certain substances in their body. They can be poisoned by medicine that is meant to help them.

Genetic testing could show whether people have genetic variants that code for these enzymes. Doctors think this can help us match medicines to patients and their genomes.

In addition, genetic testing could also reveal that a person has alleles that can lead to disease. In these cases, treatment could be started early. This could include medicines and lifestyle changes, and could be very useful for diseases such as cardiovascular disease. Certain genetic variants can increase a person's risk of developing this kind of disease, but lifestyle is also very important. You will learn more about these kinds of diseases in B2.

These two uses of genetic testing are parts of what doctors call **personalised medicine**.

Genetic screening

Personalised medicine requires large numbers of people to undergo **genome sequencing**. In whole-genome sequencing, the complete sequence of bases in an individual's DNA is worked out. Patterns in the data can then be linked to their medical histories and use of medicines. Machines analyse thousands of samples of the individual's DNA, and the genome sequence is compiled by a computer. In 2015 this would take a few days but the process is getting faster all the time.

Find out about
- how our increasing understanding of the human genome could affect healthcare
- personalised medicine as an application of gene technology
- what needs to be considered when using gene technology in healthcare

Key words
➤ personalised medicine
➤ genome sequencing

Nisha had a dangerous reaction to a medicine. Genetic testing could help to avoid this.

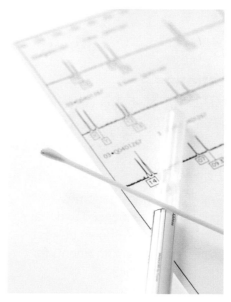
Decisions about healthcare could be based on genetic test results.

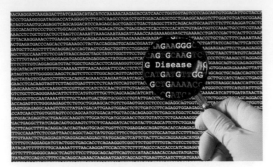

Genomics England is sequencing the genomes of hundreds of thousands of people in England.

There are plans to fully sequence the genomes of 100 000 people in England by 2017. This will be done by a project called *Genomics England,* which is funded by the UK Government.

Should we have a programme of genetic screening for everybody? It is easy to see why people may want genetic screening:

- People would know the risks of passing harmful genetic variants to their children, and could make informed decisions about family planning.
- People would know if they were at risk of developing diseases later in life, and could start treatment early.

Is it right to use genetic screening?

At first glance, genetic screening seems like the best course of action for everyone. It may benefit many people. But it may not be the right decision if it causes a great amount of harm to a few people. There are ethical and practical questions to consider about genetic screening.

Questions

1 Write down two ways in which genetic testing can be used to provide personalised medicine.

2 What is genome sequencing?

3 Some people think genomes should not be sequenced and that health information should be kept private. Write down arguments for and against this idea.

The right to privacy

Once the result of a genetic test is known, it cannot be undone and it may be difficult to keep it secret. Who else should know? Your family? An employer? An insurance company? Many people think only you and your doctor should know information about your genome. They are worried it could affect a person's job prospects and life insurance.

People with life insurance pay a regular sum of money to an insurance company. When they die, the insurance company pays out an agreed sum of money. People buy life insurance to support their families when they die. Since 2001, UK insurance companies do not require the results of a genetic test when a person buys life insurance. This agreement will be reviewed in 2017. The one exception is Huntington's disease, where test results do need to be declared.

C: Genetic engineering

Gene technology could help us feed the growing population of the world and keep more people healthy. **Genetic engineering** is an example of this new technology. It involves modifying the genome of an organism to introduce one or more desirable characteristics.

In Hawaii, 75% of the papaya grown is genetically engineered to be resistant to ringspot virus. But this fruit cannot be imported or eaten in Europe.

Genetically engineered bacteria can make medicines such as insulin (used to treat the disease diabetes). We can introduce resistance to the ringspot virus into papaya plants, to protect this food source and the livelihoods of farmers in Hawaii. And we can make genetically engineered 'golden rice'. This rice contains beta-carotene to boost vitamin A levels in millions of people who eat rice around the world.

Putting new genes into cells

One example of genetic engineering is adding one or more genes into a genome. DNA that codes for the production of insulin or beta-carotene can be taken from one organism and added into the genome of another. This works because all organisms have DNA and use the same bases (ACTG) in their genetic code.

The steps in the process are:

1 Prepare the DNA by isolating the gene from the donor organism and copying it many times.

2 Put the gene into a **vector** that will carry the gene into the target cell.

3 Select the cells that have successfully taken up the DNA and grow them further.

Genetic engineering of bacteria

A vector is needed to carry one or more new genes into a cell. To modify bacteria, scientists often use plasmids as vectors. Bacteria take up plasmids into their cytoplasm and use the genes on the plasmid. As the bacteria reproduce, each new bacterium contains a copy of the plasmid.

Not all the cells in a population of bacteria will take in the plasmid vector. Scientists put a second gene into the plasmid to make the genetically modified cells easy to select. For example, there is a gene in jellyfish that codes for a green fluorescent protein and several genes make bacteria resistant to antibiotics.

Find out about

- the process of genetic engineering
- how desirable characteristics can be introduced into an organism
- what needs to be considered when using genetic engineering in agriculture

Key words

➤ genetic engineering

H

➤ vector

Golden rice (left) is genetically engineered to contain beta-carotene, which the human body uses to make vitamin A.

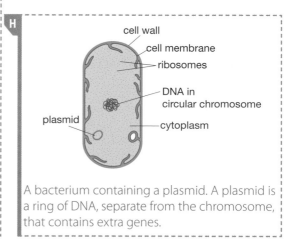

A bacterium containing a plasmid. A plasmid is a ring of DNA, separate from the chromosome, that contains extra genes.

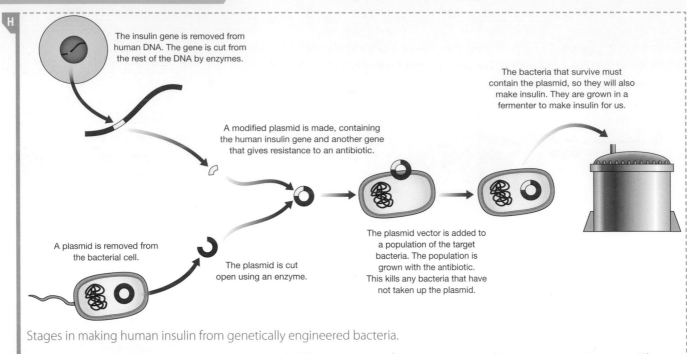

The insulin gene is removed from human DNA. The gene is cut from the rest of the DNA by enzymes.

A modified plasmid is made, containing the human insulin gene and another gene that gives resistance to an antibiotic.

The bacteria that survive must contain the plasmid, so they will also make insulin. They are grown in a fermenter to make insulin for us.

A plasmid is removed from the bacterial cell.

The plasmid is cut open using an enzyme.

The plasmid vector is added to a population of the target bacteria. The population is grown with the antibiotic. This kills any bacteria that have not taken up the plasmid.

Stages in making human insulin from genetically engineered bacteria.

Several different types of enzyme are used in genetic engineering. They recognise particular sequences of nucleotides in DNA. The first type cuts the DNA when it finds the correct sequence. This is how a gene can be isolated from DNA. The second type joins nucleotides together. This one is used to insert genes into a plasmid.

An organism that has been modified to include genetic material from another organism is a **transgenic** organism.

Genetic engineering of plants

In developing countries, about 40% of children under the age of five suffer from a lack of vitamin A. This causes blindness and reduces the ability to fight infections. About 2.7 million children die every year as a result.

Golden rice was made to contain beta-carotene, which our bodies can use to make vitamin A. Three genes were introduced into the rice plant using a plasmid and a bacterium called *Agrobacterium* as vectors. The modified plasmid was added to an *Agrobacterium* cell, which then infected a cell from a rice plant.

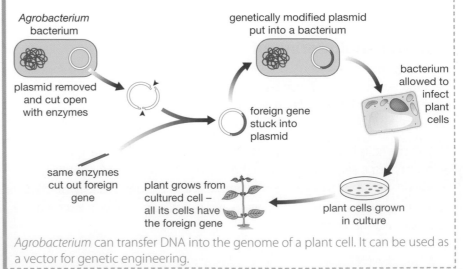

Agrobacterium bacterium

plasmid removed and cut open with enzymes

same enzymes cut out foreign gene

foreign gene stuck into plasmid

genetically modified plasmid put into a bacterium

bacterium allowed to infect plant cells

plant cells grown in culture

plant grows from cultured cell – all its cells have the foreign gene

Agrobacterium can transfer DNA into the genome of a plant cell. It can be used as a vector for genetic engineering.

Key word

➤ transgenic

Genetic engineering has been used to make crops with improved resistance to pests and diseases. This increases yields and reduces the need to spray the crops with pesticides. These crops are sometimes called genetically modified (GM) crops.

Why are some people concerned?

Some people are worried about what might happen if GM crops breed with non-GM plants. They are concerned that inserted genes could spread to other species. They also suggest that humans and other animals could experience harmful effects when they eat GM crops.

Careful tests on golden rice show it to be nutritious and safe to eat. The risk of allergic reactions to the food is very small. The risk of the inserted genes spreading from golden rice to non-GM crops is low and manageable. The scientist who created golden rice has made sure that it is given freely to farmers in developing countries. These farmers are allowed to keep and replant the seeds.

Even so, not everybody agrees with creating and using GM crops. In 2013, protesters destroyed an experimental field of golden rice being grown in the Philippines. In countries such as the UK and the USA, many people object to genetic engineering because they believe it is unethical.

The potential risks and benefits of GM crops need to be weighed up. Scientists must communicate their work to the public and to governments, as well as to each other. This helps everybody to make informed decisions about issues such as genetic engineering.

You should be careful when you read about GM crops. A lot of the information you might find is produced by people who feel very strongly about the issues. They might focus on the benefits to promote the development of GM crops. Or they might focus strongly on the risks to discourage the use of genetic engineering. It is important to know who has produced each piece of information. This helps you to decide how it fits into the argument. Not every source of information will give a balanced view of the issues.

Many crops are treated with pesticides to reduce pest damage and disease. This can cause pollution and damage ecosystems. An alternative solution is to grow crops that have been genetically engineered to be resistant to pests and disease.

Long-term trials of genetically engineered crops are important. But they can be a target for protests.

Questions

1 Draw a labelled diagram of a bacterial cell. Highlight the parts that contain genetic information.

2 What is a plasmid?

3 Make a flow chart to show the main stages in genetically engineering bacteria. Highlight the vector in the process.

4 Explain why people may be more willing to use genetically engineered insulin made in a fermenter than they are to eat golden rice?

Science explanations

B1 You and your genes

In this chapter you learnt how the genome and the environment affect an organism's features, how genetic information is inherited, the benefits of some different types of gene technology, and issues that arise from these technologies.

Pairs of chromosomes in a human genome.

You should know:

- that the genome is the entire genetic material of an organism and a copy of the genome is present in every cell to control how it functions

- that the genome is packaged into chromosomes, which are made of DNA – a polymer of nucleotides, forming two strands in a double helix

- that genes are sections of DNA, and instruct cells how to make proteins from amino acids

- that most of an organism's features depend on instructions in the genome and can be modified by the environment

- that the two versions of each gene in a pair of chromosomes are called alleles, which can be the same (homozygous) or different (heterozygous), and that a different version of a gene is a genetic variant

- that the genotype of an organism is the genetic information it has, while the phenotype is the feature or features that result from this information and interaction with the environment

- about dominant and recessive alleles, and that the recessive allele in a heterozygous pair will not show its associated feature in the phenotype

- how to use simple models (Punnett squares and family trees) to explain and make predictions about the inheritance of single genes

- that most of an organism's features are affected by multiple (rather than single) genes, other regions of the genome, and the environment

- how sex is determined by the sex chromosomes

- how gene technologies such as genetic testing, pre-implantation genetic diagnosis, and genetic engineering could be helpful in medicine and agriculture

- about risks, ethical issues, and practical issues that arise from gene technologies.

Ideas about Science

Models are used in science to help explain ideas and make predictions about possible outcomes. Punnett squares and family trees are models of inheritance. For both of these models, you should be able to:

- describe which bits of information and data are included in the model
- use the model to explain or make a prediction about inheritance
- identify limitations of the model.

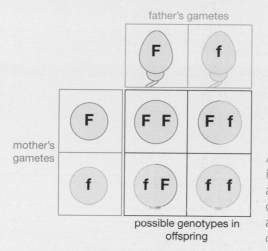

father's gametes

mother's gametes

possible genotypes in offspring

A Punnett square is used to model inheritance. In this case, both parents are carriers of the genetic variant (f) that causes cystic fibrosis. If the couple have a child, the probability that it will have cystic fibrosis is 0.25.

Our developing understanding of the genome is bringing with it new gene technologies such as genome sequencing, genetic testing, pre-implantation genetic diagnosis, and genetic engineering. These applications of science could help us to improve healthcare. They could also help us to feed the growing world population.

However, new technologies can introduce risks. The size of a risk depends on the chance of it happening and the consequences if it did. For technologies such as genetic testing and genetic engineering, you should be able to:

- identify risks and benefits to the different people involved
- discuss a course of action, taking account of who benefits and who takes the risks
- suggest reasons for people's willingness to accept the risk.

We must also consider ethical questions about how and whether gene technologies should be used. These are questions about right and wrong, and cannot be answered by science. Different people may have different views. You should be able to:

- identify ethical issues
- summarise different views that may be held
- suggest why different decisions may be appropriate in different personal, social, and economic contexts.

Scientists must communicate their work to a range of audiences, including the public, other scientists, and politicians, in ways that can be understood. This helps people to make informed decisions about the use of new technologies.

Genome sequencing and genetic testing have great potential benefits. But they also have risks, and there are ethical questions about how and whether we should use them.

B1 Review questions

1 Which statement describes a genome?

   A The nucleus of a plant or animal cell.

   B A section of DNA that describes how to make a protein.

   C All the genes an organism inherits.

   D The entire genetic material of an organism.

2 Read the following statements.

   A Different versions of a gene, caused by small changes in DNA.

   B The combination of alleles that an organism has.

   C The human genome is organised into 23 pairs of these.

   D The features that an organism has.

   a Which statement describes chromosomes?

   H **b** Which statement describes genetic variants?

   c Which statement describes the genotype?

3 Cystic fibrosis (CF) is a genetic disease.

   a Write down the word from the list that describes the genetic variant that causes CF:

   dominant   homozygous
   heterozygous   recessive

   b The family tree shows the inheritance of CF.

   i Which person, **A**, **B**, **C**, or **D**, is a female who has inherited two faulty CF alleles?

   ii Which people from **A**, **B**, **C**, and **D** are carriers?

   iii Person **B** has a daughter. We cannot tell from the family tree whether the daughter is a carrier. Explain why.

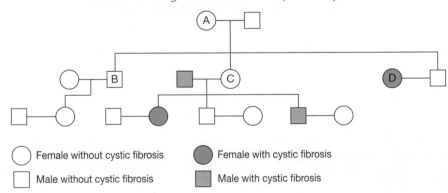

   ○ Female without cystic fibrosis   ● Female with cystic fibrosis

   ☐ Male without cystic fibrosis   ■ Male with cystic fibrosis

4 Some people have short fingers. This feature is caused by a single gene. There are two genetic variants of this gene. The dominant B variant causes short fingers. The recessive b variant causes normal fingers.

a Copy and complete the Punnett square to show all the possible genotypes in the children.

b What proportion of the children are likely to inherit a B variant?

A 25%   B 50%   C 75%   D 100%

c What is the probability that a child will have a heterozygous genotype (Bb or bB)?

d What will be the likely ratio of short-fingered children to normal-fingered children?

gametes	B	b
B		
b		

5 Science can show how genetic testing can be carried out. It cannot explain whether it should be carried out.

a Describe some implications of carrying out genetic testing on human beings.

b Explain the ethical issues involved in genetic testing and the different views that might be held.

6 Imagine you had to approve the growth of genetically engineered 'golden rice' in this country. What risks and benefits would you need to think about?

7 The diagram shows the genetic engineering of a bacterium to produce human insulin.

a What is genetic engineering?

b In the diagram, what is the vector?

c What does the vector do in this process?

d How would you select modified bacteria?

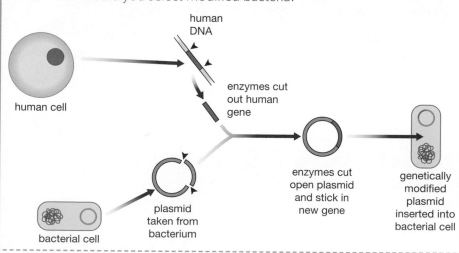

human cell

human DNA

enzymes cut out human gene

bacterial cell

plasmid taken from bacterium

enzymes cut open plasmid and stick in new gene

genetically modified plasmid inserted into bacterial cell

B2 Keeping healthy

Why study health and disease?

Good health is something everyone wants. Stories about keeping healthy are all around you, for example, news reports about what to eat, how much to drink, new viruses, and 'superbugs'. New evidence is reported every day. The message about how to stay healthy often changes. It's not always easy to know which advice is best.

What you already know

- Good hygiene helps humans keep healthy.
- What the main parts of the human circulatory system do, including the heart, blood vessels, and blood.
- Bacteria are an important part of the human digestive system.
- Animals, including humans, need the right types and amount of nutrition. A healthy human diet includes carbohydrates, lipids (fats and oils), proteins, vitamins, minerals, dietary fibre, and water.
- Obesity, starvation, and certain diseases can occur when there are imbalances in the diet.
- Diet, exercise, drugs, and lifestyle affect the way our bodies work.
- Misusing drugs can affect our behaviour, health, and life processes.
- The information in your genome can affect your characteristics, including your health.

The Science

Some diseases are caused by harmful microorganisms. If you are infected, your body has amazing ways of fighting back. Vaccines and drugs can help you survive many diseases, and doctors are always trying to develop new ones. But not all diseases are caused by microorganisms. Your lifestyle may also put you at risk of disease. Media reports often warn about the dangers of smoking, eating badly, and not exercising.

Ideas about Science

How do you decide which health reports to trust? It's helpful to know about correlation and cause, and about how scientists check each other's work. There are also ethical questions (about right and wrong) to consider when deciding how we should use vaccines and drugs.

B2.1 What are the causes of disease?

A: In the doctor's waiting room

Find out about

- the relationship between health and disease
- what causes the diseases you can catch
- how bacteria reproduce

When you feel fit and well, you probably take your good health for granted. But when you're ill, you realise how important your health is. Disease can cause problems with your physical and mental health. Some diseases – such as a cold – are minor and you recover quickly. Other diseases – such as heart disease and cancer – can be life-threatening.

You might go to see a doctor if you have **symptoms**. Symptoms are the feelings or changes you experience in your body when you are ill. You can get symptoms when your body's cells or systems have been damaged or are not working normally.

runny nose, sore throat, aching muscles, fever, and swollen glands

skin between the toes is itchy, red, and flaky

cut finger which is red, sore, and filled with pus

symptoms of a sexually transmitted infection

chest pains when exercising

In the doctor's waiting room different people have different symptoms.

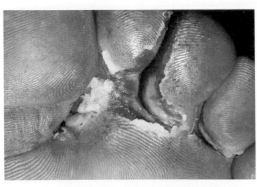

The fungus that causes athlete's foot grows on the skin.

Can you catch it?

Some diseases are caused by **viruses**, **bacteria**, **fungi**, or **protists** that infect the body. A disease caused by an infection can be passed from one organism to another. It is a **communicable disease**.

In the doctor's waiting room, Louise has itchy, red, flaky skin between her toes. She has a disease called athlete's foot, which is caused by a fungus. She caught the disease when the fungus was transferred to her skin from a contaminated surface.

Other diseases are caused by a person's lifestyle, their environment, or their genes. This kind of disease is not caused by an infection. It is a **non-communicable disease**.

In the doctor's waiting room, Oliver feels chest pains when he exercises. This is a symptom of coronary heart disease. A person can develop this disease if they have an unhealthy lifestyle. You will learn more about non-communicable diseases later in this chapter.

What are viruses, bacteria, fungi, and protists?

Bacteria are single-celled organisms. They are prokaryotic organisms, which means their cells do not have a nucleus. The DNA in a bacterium is found in the cell cytoplasm.

Every surface you touch is covered with bacteria. They also live on the outside and inside of your body. Most bacteria do not cause human diseases. Some can even be helpful, for example, bacteria that live in your gut help you to digest food and protect you against harmful bacteria.

Antibacterial gel kills harmful bacteria on your skin. Yoghurts containing live bacteria top up the populations of helpful bacteria that live in your gut.

Fungi and protists are eukaryotic organisms, which means their DNA is packed inside a nucleus in each cell. Most fungi are multicellular, but some (such as yeasts) are single-celled organisms. Protists can be single-celled or multicellular.

Bacteria, fungi, and protists can reproduce whenever the conditions are warm with plenty of nutrients, moisture, and space.

Viruses are very different. They are not made of cells. Each virus particle is made of genetic material wrapped in a layer of protein and sometimes fat molecules. Scientists have debated for many years whether or not viruses are alive. Viruses can only reproduce by invading living cells.

A bacterium, fungus, protist, or virus that causes disease is a **pathogen**.

Key words

- ➤ symptom
- ➤ virus
- ➤ bacterium
- ➤ fungus
- ➤ protist
- ➤ communicable disease
- ➤ non-communicable disease
- ➤ pathogen

	Width (µm)
Virus	0.02–0.3
Bacterium	1–5
Protist	1+
Fungus	50+
Human hair	17–181

How small are pathogens compared to a human hair? One micrometre is one thousandth of a millimetre. Viruses are too small to be seen, even using a light microscope.

Invasion of the pathogens!

In the doctor's waiting room, Jolene has a sore finger. She cut it when she was gardening. Bacteria on her skin and in the soil invaded the cut. Once inside they started to reproduce.

Reproduction in bacteria is simple. Each bacterium divides to produce two new ones. These grow for a short time before dividing again. In ideal conditions, such as inside the human body, bacteria can divide once every 20 minutes.

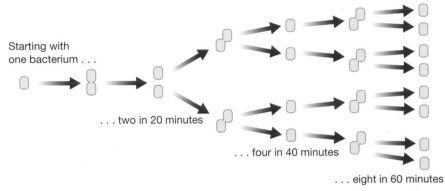

Starting with one bacterium . . .

. . . two in 20 minutes

. . . four in 40 minutes

. . . eight in 60 minutes

Bacteria can reproduce quickly inside the body.

Worked example: Reproducing bacteria

One hundred bacteria are grown in ideal conditions. Each bacterium divides once every 20 minutes. How many bacteria will there be after one hour?

Step 1: Write down what you know, with the units.

number of bacteria at start = 100
end time = 1 hour
number of bacteria will double every 20 minutes

Step 2: It's easier to work out the answer if all of the times are in minutes. So convert the end time from one hour into minutes. There are 60 minutes in each hour.

end time = 1 × 60 minutes
= 60 minutes

Step 3: Make a table with time (in minutes) in one column, and the number of bacteria in another column. Each row represents one 20-minute cycle of the bacteria dividing.

Start at zero minutes.

Work down each row in the table. In each row:

- add 20 minutes to the time
- double the number of bacteria.

Time (minutes)	Number of bacteria
0	100
20	200
40	400
60	800

Answer:

The table shows that after 60 minutes (which is the same as 1 hour) there will be 800 bacteria.

Bacteria can't keep dividing forever. When nutrients and space start to run out, they limit the rate at which bacteria reproduce. Also, the bacteria's own waste products can eventually poison the bacteria and kill them.

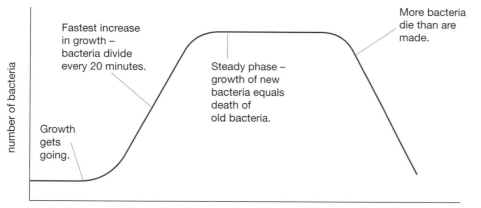

Bacteria in a sealed container can't keep reproducing at their fastest rate. Their food starts to run out, or waste products kill them off.

Causing symptoms

Pathogens can cause symptoms of disease when they:

- damage our cells by invading and reproducing
- release toxins (poisons).

Symptoms may only appear after a pathogen has divided enough times to produce a large number of pathogens in the body. This delay between infection and the appearance of symptoms is called the **incubation period**.

The incubation period can be anything from a few hours to several years, depending on the pathogen. A person who has an infection but is still in the incubation period can appear to be healthy because they are not yet showing symptoms.

Questions

1 Write down three types of organism that can cause disease.

2 Why is it important to clean your hands with antibacterial gel when visiting a sick relative in hospital?

3 Explain why a person may appear healthy even when they have an infection.

4 Three bacteria get into a cut. How many might there be after three hours in ideal conditions?

B: Passing it on

- how communicable diseases spread
- some common communicable diseases in humans and in plants

Pathogens can be passed from an infected organism to a healthy one. This is how a communicable disease spreads.

Pathogens can be passed between organisms:

- in body fluids (including blood, saliva, mucus, and semen) and waste products (including faeces)
- in contaminated food and water
- by touch.

Some common human diseases show how different pathogens spread and cause disease.

Disease	Pathogen	Spread:	Infects a human when:
Flu	influenza viruses	in droplets of body fluids (from coughs and sneezes) in the air and on surfaces	droplets are inhaled or transferred from hand to nose, mouth, or eye
***Salmonella* food poisoning**	*Salmonella* bacterium	in contaminated food and water	contaminated food or water is ingested
Athlete's foot	fungus	on surfaces (e.g., changing room floors)	skin touches a contaminated surface
Malaria	*Plasmodium* protists (purple)	by mosquitoes	mosquito bite introduces infected saliva into the human's blood

The images of these pathogens were made using electron microscopes. Colour has been added artificially by a computer.

Sexually transmitted infections

Some diseases are spread by unprotected sex or genital contact. Unprotected sex means having sex without using a condom. A disease that is spread this way is called a **sexually transmitted infection (STI)**.

There are many common STIs in the UK. Chlamydia and gonorrhoea are caused by bacteria. Genital warts are caused by viruses.

Acquired immune deficiency syndrome (**AIDS**) is an STI that kills millions of people worldwide each year. It is caused by human immunodeficiency virus (**HIV**). The virus is spread in semen and blood. It is usually transmitted during unprotected sex. It can also be passed on by sharing needles used to inject drugs. There is a very long incubation period for HIV. A person infected with HIV will not usually show the symptoms of AIDS until many years later. AIDS severely damages the body's ability to fight off other infections, and there is no cure.

It is important to use condoms when having sex to reduce the risk of catching STIs.

Plants get diseases too

It's not just humans that catch diseases. Other animals and plants catch diseases too. There are even viruses that attack bacteria and fungi.

Plant and animal diseases can damage ecosystems and food chains. The loss of crops and farm animals can disrupt our supply of food, and has economic consequences.

Crown gall disease

Pathogen: bacterium

Infects: grapes, stone fruit and nut trees, and sugar beet

Spread by: contaminated soil, water, and farming tools

Crown gall is one of the most serious plant diseases. The pathogen infects over 1000 different plant species worldwide.

Ash dieback

Pathogen: fungus

Infects: ash trees

Spread by: spores carried by wind, and movement of contaminated plant material

Found across Europe, and in the UK since 2012. Can kill up to 90% of ash trees in a country. Most countries that have tried to control the spread of the disease have failed.

Tobacco mosaic virus

Pathogen: virus

Infects: tobacco, tomato, bell pepper, cucumber, some flowers

Spread by: direct contact and contaminated seeds

Tobacco mosaic virus (TMV) was the first virus to be discovered. By studying TMV, scientists have learnt about viruses and how they interact with plants.

Notes about some common plant diseases. Tobacco mosaic virus is studied by scientists as a **model organism**.

It is important to reduce the spread of diseases in plants, humans, and other animals. You will learn about ways of doing this later in this chapter.

Key words

➤ sexually transmitted infection (STI)
➤ AIDS
➤ HIV
➤ model organism

Many of the world's staple foods come from crops that can be attacked by diseases.

The brown patch shows that this tree is infected with the fungus that causes ash dieback.

Questions

1 Write down one crop that is threatened by tobacco mosaic virus.

2 Suggest why some people wear flip-flops on their feet when showering at the swimming pool.

3 Explain why you cannot catch malaria by hugging a person with the disease.

4 Explain why the destruction of a crop plant by a disease can have economic consequences.

B2.2 How do organisms protect themselves against pathogens?

A: Keep out!

Find out about

- physical, chemical, and bacterial defences against pathogens in humans
- how platelets are adapted to seal wounds

Key words

➤ non-specific defences
➤ platelets

> It was just a small cut, so I ignored it. By the time I went to bed it was a bit sore and red. Now it's swollen and shiny. It really hurts.

Jolene is worried about her cut finger.

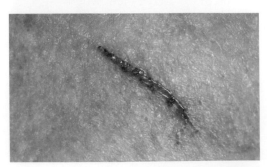

Your skin stops pathogens entering your body. But they may be able to get in through a cut.

There are pathogens all around you, all the time. They're on surfaces, in the air, on food, and in water. You are able to stay healthy most of the time because your body has defences against infection.

Human defences

Our bodies have defences that make it difficult for pathogens to enter the blood. These defences include physical barriers, chemical substances, and even helpful bacteria.

Type of defence	Examples	How do they help?
physical defences	the skin; mucus in the respiratory and digestive systems; platelets in the blood	prevent pathogens from invading our tissues
chemical defences	stomach acid; enzymes and substances that kill microbes in saliva, mucus, and tears	destroy pathogens and stop them reproducing
bacterial defences	helpful bacteria that live on the skin and in the gut	compete against pathogens by using up nutrients and space

These defences are always present. They are not made in response to a specific pathogen. They are **non-specific defences**.

Sealing the wound

Remember the doctor's waiting room from B2.1A? Jolene had a cut finger. Her skin was cut open, allowing bacteria to enter her body.

Blood vessels in Jolene's skin were cut. Blood contains **platelets** that help to seal wounds. Platelets are fragments of cells made from the cytoplasm of large cells. When a blood vessel is damaged, platelets stick to the cut edge. They send out substances that trigger a series of reactions. This makes blood clot at the cut site.

Blood clotting helps prevent pathogens from entering blood vessels. It also stops you losing too much blood.

Questions

1 Name two places in the human body where chemical defences against pathogens are found.

2 What type of defence was damaged when Jolene cut her finger?

3 Suggest what might happen to Jolene's cut if she didn't have platelets.

B: Fighting back

Some pathogens manage to get past our non-specific defences and get inside our bodies. It is the job of our **immune system** to fight these pathogens. Our species would have died out long ago if the immune system had not evolved.

The battle for Jolene's finger

The area around the cut on Jolene's finger is red, swollen, and filled with pus. The redness and swelling is inflammation. Extra blood is being sent to the wounded area, bringing the immune system's main defenders – **white blood cells**. Their function is to destroy pathogens.

One type of white blood cell ingests bacteria (takes them inside itself) and then digests them. A different type of white blood cell makes **antibodies**. These are molecules that stick to specific pathogens. Antibodies can disable pathogens and label them for attack by other white blood cells.

Worn-out white blood cells, dead bacteria, and broken cells collect as pus. As the bacteria are killed, the amount of inflammation and pus decreases until the tissue heals.

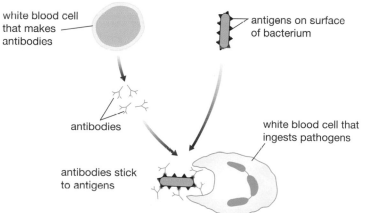

A different type of white blood cell makes antibodies that stick to specific pathogens. (Note: the white blood cells, antibodies, and bacteria have not been drawn to scale. White blood cells are many times larger than bacteria, and bacteria are many times larger than antibodies.)

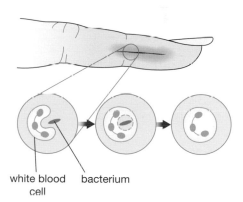

One type of white blood cell ingests and then digests bacteria.

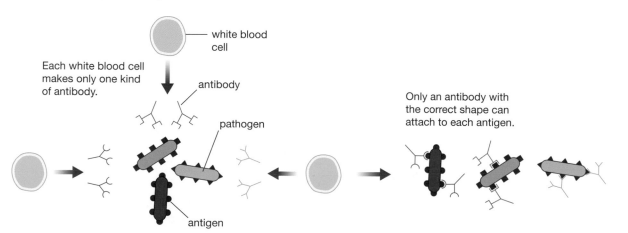

A different antibody is needed for each different pathogen. (Note: the white blood cells, antibodies, and bacteria have not been drawn to scale.)

Recognising pathogens

How do Jolene's white blood cells recognise what to attack? It would be very dangerous if they attacked her body cells.

All cells and pathogens have molecules on their surface called **antigens**. White blood cells have receptors on their surface that recognise antigens. The receptors are molecules that have the correct shape to recognise specific antigens.

The antigens on body cells are recognised as 'self'. This means white blood cells do not attack. But the antigens on pathogens are recognised as 'non-self'. This triggers an immune response.

The antigens on every pathogen are unique. Each white blood cell makes only one type of antibody, which recognises a specific antigen. Your body has to make different antibodies for different pathogens. Making antibodies against a new pathogen takes a few days, so you can get symptoms before your white blood cells destroy the pathogen.

Key words

➤ antigens
➤ immunity
➤ memory cell

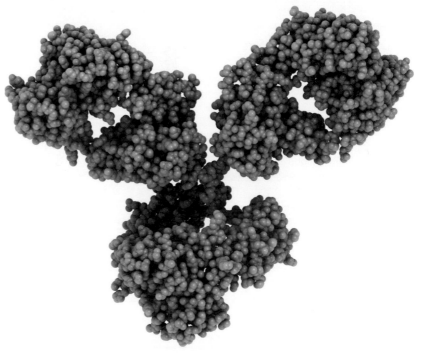

A computer model of a single antibody. Each Y-shaped antibody molecule has two 'arms' (top) that stick to the antigen. Antibodies are made of protein.

Why do I get some diseases just once?

You might have heard that if you catch chickenpox when you're young you can't catch it again. Your body has developed **immunity** to it. But how does this work?

After you have been infected with the chickenpox virus, some of the white blood cells that made antibodies against it stay in your blood. These are **memory cells**. If you are infected with the chickenpox virus again, the memory cells reproduce and make the antibody very quickly. So the immune system responds much faster the second time. You destroy the pathogen before it can make you ill. This works for other pathogens too.

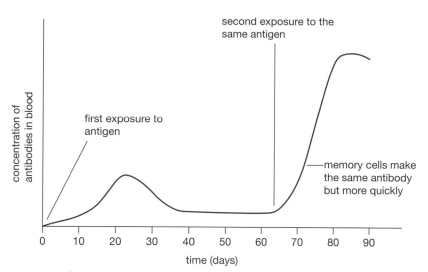

A person is infected twice by the same pathogen. Their white blood cells make antibodies much faster the second time.

Why do I get some diseases more than once?

Some diseases – such as the common cold – are caused by many different pathogens. Memory cells that recognise a particular pathogen cannot make antibodies against a different pathogen. This is why catching one cold does not make you immune to catching another.

To make things worse, the pathogens that cause some diseases – such as flu – change their antigens. This happens because of mutations in their DNA. Mutations can change the shape of the antigens. Memory cells from an earlier infection with the pathogen will not recognise the changed antigens. So you can suffer the symptoms of flu all over again.

Questions

1   Explain two ways in which white blood cells protect the body from invading bacteria.

2   Draw a flowchart to explain how you can become immune to chickenpox.

3   A person is infected with the influenza virus. Later, they are infected with the virus again but the virus antigens have changed. Sketch a graph to show how the concentration of antibodies in this person's blood changes over this period.

B2.3 How can we prevent the spread of infections?

A: Stopping the spread

Find out about

- how the spread of infections can be reduced and prevented in humans and plants
- preventing the spread of diseases in different parts of the world

Why is it important to wash your hands after using the toilet and before meals? Why do you keep some foods in the refrigerator? Why do you cook some foods before you eat them? Doing these things helps us to stay healthy by stopping the spread of pathogens.

Although there are medicines to treat many diseases, it's always better to avoid infection in the first place. If communicable diseases were allowed to spread freely, many people, other animals, and plants would die.

Healthy at home

There is a lot you can do at home to prevent the spread of infections.

1 Cleaning and covering wounds helps keep them **sterile** (free of bacteria, fungi, protists, and viruses).

2 Clean water is free of pathogens.

3 Effective **sanitation** safely removes faeces and waste, which can carry pathogens.

4 Good personal **hygiene** removes pathogens (and substances that trap them) from the skin.

5 Keeping food cold in the refrigerator reduces the rate at which bacteria and fungi reproduce.

6 Preparing food hygienically, using clean utensils and surfaces, avoids **contamination** with pathogens.

7 Cooking food for long enough at high temperature kills bacteria and fungi.

Bacteria, fungi, protists, and viruses can be killed by cleaning products, including bleach.

Food poisoning is caused by consuming food contaminated with bacteria, fungi, or viruses. It can be fatal. Bacteria and fungi can be killed by cooking, but they can release toxins (poisons) into food that are not destroyed by cooking. Some bacteria form **spores** that survive cooking and then start growing again. So it's important to avoid contamination of food during storage and preparation.

UK law includes food hygiene regulations for businesses that handle foods. Environmental Health Officers can close down a food manufacturer, shop, or restaurant if it does not follow the regulations.

Key words

- sterile
- sanitation
- hygiene
- contamination
- spore
- epidemic
- pandemic

Stop it!

Every cough and sneeze sends millions of droplets of saliva and mucus into the air. Each droplet contains pathogens. Coughing and sneezing into a tissue stops the pathogens spreading.

A high-speed snapshot of a sneeze.

In some cases the spread of a disease could pose a serious threat to public health. It may be necessary to stop people travelling into and out of an area. This happened with an outbreak of Ebola disease in West Africa that started in late 2013.

If a disease affects many individuals in a population, this is an **epidemic**. If it spreads over a very wide geographical area it is a **pandemic**.

Sneezing and coughing into a tissue is an easy way to stop the spread of the pathogens that cause colds and flu.

CATCH IT
Germs spread easily. Always carry tissues and use them to catch your cough or sneeze.

BIN IT
Germs can live for several hours on tissues. Dispose of your tissue as soon as possible.

KILL IT
Hands can transfer germs to every surface you touch. Clean your hands as soon as you can.

NHS

A poster from the UK National Health Service on how to avoid spreading colds and flu.

Key words

➤ vaccination
➤ contraception
➤ ecosystem service
➤ monoculture

Synoptic link

You can learn more about vaccines in B2.3B *Vaccines*.

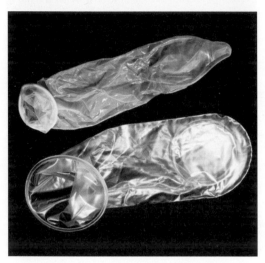

A male condom (top) is worn over the erect penis. A female condom (bottom) is worn inside the vagina.

Vaccinations

You've probably already had a **vaccination**. It usually involves an injection. Vaccinations have been used very effectively to reduce the spread of infections.

Other animals

It's important to prevent the spread of infections in animals other than humans. We depend on many animals for food. Animals that we care for must live healthy lives. Good hygiene and sanitation are important when handling animals. Animals can have vaccinations too. But when there is an outbreak of a disease, it may be necessary to kill infected animals to prevent an epidemic.

Contraception

The spread of STIs, including HIV/AIDS, can be reduced by using condoms when having sex. Condoms are a form of **contraception**. They prevent body fluids from one person entering the other during sex.

Around the world

In many places, people live in poverty. Sanitation is poor, conditions are overcrowded, and there is a lack of clean water. It can be difficult to access contraception and healthcare. Diarrhoea, malaria, and AIDS are big killers.

The World Health Organization (WHO) reports that in low-income countries, 40% of all people who die are children under 15 years old (it is 1% in high-income countries). Most of these deaths are caused by infections, so they could be prevented.

The United Nations is working to reduce poverty and fight the spread of diseases. International efforts have improved living conditions, provided health information, and rolled out vaccination programmes in developing countries. But there is still much work to be done.

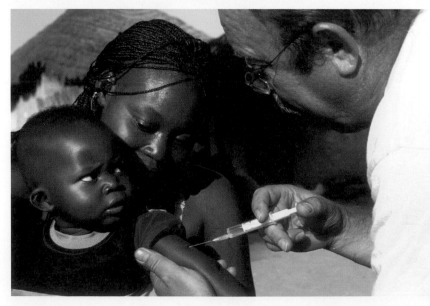

An infant is vaccinated at a health clinic in Senegal, West Africa.

Protecting plants

Every food chain starts with a producer such as a plant. They create habitats for many organisms to live in. Plants provide many **ecosystem services** that benefit humans. So it's very important to try to prevent the spread of diseases in plants.

Regulating the movement of plant material

In most countries, farmers have been growing crops and managing woodland for centuries. They have learnt how to keep them healthy. Because of natural selection and selective breeding, plants growing in a particular place are usually well adapted to survive there. The biggest threat to plant health is when infected plants or seeds are brought in from other areas or countries.

It is very important to regulate and control the movement of plant material. This includes foods, seeds, timber, whole plants, soil, and so on. All plant material is inspected when it enters the UK via a dock or an airport. If evidence of infection is found, the plant material will be safely disposed of. When buying plants or seeds to grow, it is important to make sure they are free of infection.

Monoculture versus polyculture

When lots of plants of the same kind are planted close together, an infection can spread quickly. This kind of planting is called **monoculture**.

Usually there is variation between individuals within a population – meaning that some will be better able to survive infection than others. But some of the crops we grow are clones, which means all the individuals are genetically identical. For example, banana plants are grown from cuttings, not seeds, so they all have exactly the same genetic variants. If an infection kills one plant, it could kill them all.

A health inspector takes samples from imported apples. The DNA of the samples is tested to look for genetic variants from pathogens.

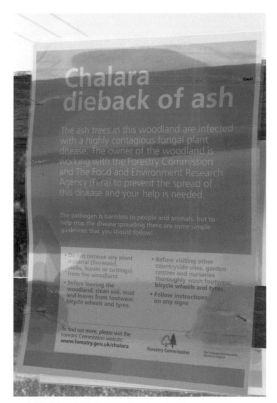

A sign on a gate in the countryside advises people on how to avoid spreading ash dieback disease.

All banana plants are clones – they have exactly the same genome. A disease that kills one plant could kill them all.

Key words

➤ polyculture
➤ crop rotation
➤ chemical control
➤ biological control

Plants grown in monoculture are easy to harvest. But growing a mixture of different plants together can help to prevent infections spreading so quickly. This is called **polyculture**. Some plants such as yarrow and calendula even help to repel insects that can carry pathogens.

Crop rotation

After a crop is harvested, pathogens or their spores can stay in the soil. If the same crop is planted again the following year, it may be re-infected. So another useful strategy is **crop rotation**. Each field is planted with a different crop each year. Not all crops are affected by the same pathogens. In addition, different crops cycle nutrients through the ecosystem in different ways. So this helps to keep the soil fertile, creating good growing conditions to encourage healthy plants.

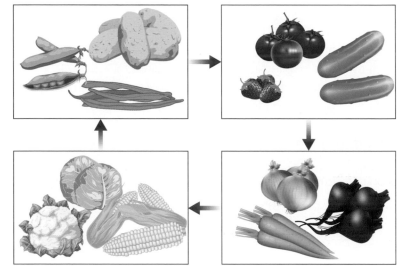

Crop rotation in four fields over four years. It's best if each field receives a different type of crop each year, for example, a legume (e.g., peas), then a fruit (e.g., strawberries), then a root vegetable (e.g., carrots), and then a leafy vegetable (e.g., cabbage).

Chemical and biological control

Sometimes, it is necessary to use chemical substances to reduce the spread of disease. This is **chemical control**. For example, half a billion people across Asia and Africa depend on bananas for food. But banana plants can easily become infected with and be killed by a fungus called *Sigatoka*. Some banana crops are sprayed with a chemical fungicide up to 40 times every year to kill the fungus.

Another option is **biological control**. This involves introducing a new species into an ecosystem. The new species is a predator. It will eat pests such as insects that spread disease and damage plants.

However, there are risks of introducing new chemical substances and species into an ecosystem.

Questions

1 Write down three ways of reducing the risk of food poisoning.

2 Increasing the availability and use of condoms could reduce the number of deaths from AIDS. Explain why.

3 When there is an outbreak of a disease in farm animals, the government may decide to kill all the infected animals. Suggest arguments for and against this action.

4 Some people like to go on walks in the countryside. They may come into contact with ash trees infected with dieback disease. Suggest how these people could avoid spreading the infection.

Synoptic link

You can learn more about the risks of introducing new substances and species into an ecosystem in B3.4A *Organisms and their habitats*.

B: Vaccines

Having a vaccination makes you immune to a disease without having to catch it first. It involves putting a **vaccine** into your body, usually by injection. It's a small jab that makes a big difference. Vaccinating many people in a population is one of the best ways to reduce the spread of an infection.

What is a vaccine?

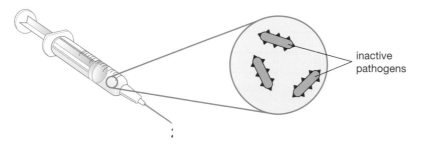

inactive pathogens

A vaccine contains dead or inactive pathogens. Sometimes just parts of pathogens are used. The vaccine makes your white blood cells respond to the pathogen's antigens but should not make you ill.

How does vaccination work?

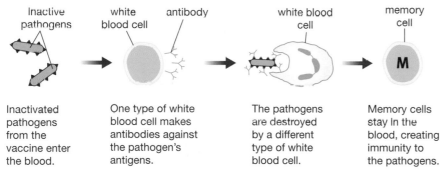

| Inactive pathogens | white blood cell | antibody | white blood cell | memory cell |

Inactivated pathogens from the vaccine enter the blood.

One type of white blood cell makes antibodies against the pathogen's antigens.

The pathogens are destroyed by a different type of white blood cell.

Memory cells stay in the blood, creating immunity to the pathogens.

The immunity you get from a vaccination should last a long time. Once you've been fully vaccinated against measles, you will be protected for at least 20 years. This works because the antigens on the measles virus do not change.

However, viruses such as influenza are another story. Their DNA mutates frequently, so they produce new antigens. The memory cells created by an old vaccination may no longer be effective against the virus with its new antigens. This is why new flu vaccines have to be made regularly.

Victory over smallpox

Smallpox was a devastating disease. In the 1950s there were 50 million cases worldwide. This fell to 10–15 million cases by 1967 because of vaccination in some countries, but 60% of the world's population was still at risk.

In 1967 WHO began a campaign to wipe out smallpox by vaccinating people across the world. The last natural case of smallpox was recorded in Somalia, in the Horn of Africa, in 1977. Smallpox was wiped out because so many people were vaccinated.

Find out about

- what a vaccine is
- how vaccination helps prevent disease
- weighing up the risks, costs, and benefits of vaccination

Key word

➤ vaccine

Smallpox killed one in every four infected people. It left most survivors with large scars and many were blinded.

The benefits of mass vaccination

In the UK there is a childhood vaccination programme against some diseases. This means that few people in the country now catch these diseases.

Age	Vaccination	Diseases protected against
2–4 months	5-in-1	diphtheria, tetanus, whooping cough, polio, and *Haemophilus influenzae* type b (bacterial infection causing meningitis and pneumonia)
	PCV	pneumococcal infections
	rotavirus	viral diarrhoea and sickness
	meningitis C	meningitis C
12–13 months	MMR	measles, mumps, and rubella
3 years	boosters	measles, mumps, rubella, diphtheria, tetanus, whooping cough, and polio
girls aged 12–13 years	HPV	cervical cancer caused by human papillomavirus
13–18 years	boosters	meningitis C, diphtheria, tetanus, and polio

The UK childhood vaccination programme.

To stop a large outbreak of a disease, almost everyone in the population needs to be vaccinated. If they are not, large amounts of the pathogen will survive in the population. It will be carried by unvaccinated, infected people. If the vaccination rate drops just a little, lots of people will become ill.

☐ vaccinated   ◼ unvaccinated   ◼ infected

98% of the population is vaccinated. Unvaccinated people are unlikely to catch the disease.

90% of the population is vaccinated. Unvaccinated people are much more likely to catch the disease.

There will always be some individuals in a population who cannot be vaccinated. This could be because they are allergic to something in the vaccine. Or it could be because their immune system is not working properly.

Everybody who can be vaccinated, should be vaccinated. When a very high percentage of a population is vaccinated, it helps protect the few people who cannot have the vaccine. This is sometimes called **herd immunity**.

Should people be forced to have vaccinations?

Governments and public bodies, such as the UK National Health Service (NHS), make decisions about who should be offered vaccinations. They do this based on an assessment of risk, cost, and benefit.

For example, here are some facts about measles and the MMR vaccination:

- Measles spreads very easily.
- Measles can be fatal (1 in 10 000 cases).
- Some children who survive measles are left severely disabled (1 in every 4000 cases).
- A vaccinated child is much less likely to get measles.
- Hardly anyone who has the vaccination notices any harmful effects.
- If experienced, harmful effects from the vaccine can be mild (3 in every 10 000 children), or can produce a serious allergic reaction (1 in every 1 million children).
- Around 1 in every 100 vaccinated children still gets measles, because vaccinations don't have a 100% success rate.
- Each vaccination costs the NHS a few pounds.
- Treating a person who has measles can cost the NHS thousands of pounds.

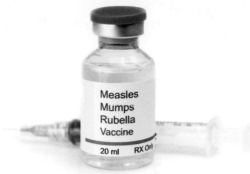

The MMR vaccine gives protection against measles, mumps, and rubella.

The NHS offers all children in the UK the MMR vaccination against measles. If everyone had to be vaccinated by law, there would be a much lower risk of any child catching the disease. There is enough MMR vaccine for every child. So the MMR vaccination could be compulsory – but it isn't.

For society as a whole, vaccination is the best choice. But society does not think it is right to force anybody to have a vaccination. Individuals have the right to decide. So there is a difference between what *can* be done with science, and what people think *should* be done.

Whose choice is it?

Vaccines are tested for safety before they are used. However, it's important to remember that no action is completely safe. People have differences in their genomes. This means they can react differently to medical treatments, including vaccines.

For each parent, deciding whether to have their child vaccinated can be a difficult choice. Parents have to balance the possible harm from the disease against the risk of possible side-effects from the vaccination. It is important that people have clear and unbiased information to help them make their decision.

Parents must decide whether to have their children vaccinated.

Research shows the MMR vaccine is safe and effective

The NHS has concluded that the MMR vaccine:

- provides effective protection against measles, mumps, and rubella
- can be safely given to children.

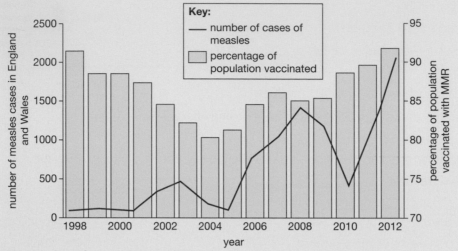

The number of people having the MMR vaccine in England fell between 1998 and 2004. There has been a general rise in the number of cases of measles since 2000. Spikes in the number of cases can be caused by new outbreaks of the disease.

It is important that as many people as possible are vaccinated. Data show that when the number of people receiving the MMR vaccination falls, cases of measles increase in the following years.

If the percentage of the population that is vaccinated stays high, the number of measles cases should eventually fall back to constant low levels.

In 1998 the MMR vaccine was wrongly linked to a disorder called autism. The disorder affects how information is processed in the brain. A study was published by Andrew Wakefield in the medical journal *The Lancet*. The study claimed there was a link between the MMR vaccine and autism. The results of the study could not be replicated by other scientists – the data were not **reproducible**. It later emerged that Wakefield had reported false evidence.

His study was declared fraudulent, and he is no longer allowed to be a doctor in the UK. However, Wakefield's study made headlines and many parents chose not to have their children vaccinated with MMR.

There was an outbreak of measles in Swansea in 2013. There were over 800 cases and one person died. Almost everybody who caught measles had not been fully vaccinated with MMR as a child. An article in the *British Medical Journal* questioned whether the outbreak was linked to reports about the MMR controversy in a Swansea local newspaper in the late 1990s.

Many scientific studies in the past decade have concluded that there is no link between the MMR vaccine and autism.

Key word

➤ reproducible

Questions

1. Describe how vaccination can stop you from catching a disease.

2. Explain why it is best if almost every individual in a population is vaccinated against a disease.

3. Make a table showing reasons for and against having the measles vaccination.

A: Cardiovascular diseases

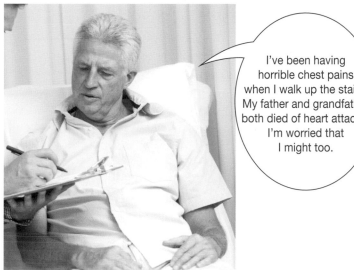

I've been having horrible chest pains when I walk up the stairs. My father and grandfather both died of heart attacks. I'm worried that I might too.

Oliver is worried about his heart.

Find out about

- risk factors for cardiovascular diseases
- correlation and cause
- health studies and sampling

Key words

- ➤ coronary heart disease
- ➤ cardiovascular disease
- ➤ coronary arteries
- ➤ heart attack

In the doctor's waiting room, Oliver is worried because he feels pains in his chest when he exercises. He has symptoms of **coronary heart disease** (CHD). This is a type of **cardiovascular disease**, a group of diseases affecting the blood vessels and heart. CHD is the UK's biggest health problem. Around 2.3 million people in the UK are living with CHD, and it kills over 73 000 people every year.

Blocking the body's supply route

Your heart beats continually. In each beat, the heart muscles contract to pump blood around your body. The blood carries food and oxygen to your cells. Cells use these substances for a supply of energy, and would die without them.

The heart muscle cells are supplied with blood by the **coronary arteries**. Sometimes fat can build up in these arteries. A blood clot can form on the fatty lump. If this blocks a coronary artery, some heart muscle is starved of food and oxygen. The muscle cells start to die. This is a **heart attack**. In the UK there are about 103 000 heart attacks every year.

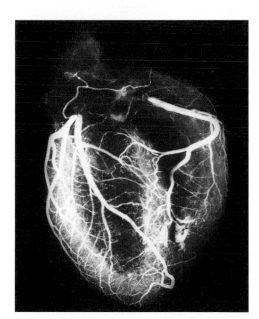

Coronary arteries carry blood to the heart muscle.

Monitoring the heart

When fatty deposits build up in the walls of the arteries, it is more difficult for the blood to be pumped around the body. The heart has to work harder and beat faster.

Fat build-up in a coronary artery.

Key words

➤ risk factors
➤ correlation
➤ outcome
➤ cause
➤ obese

You can measure how fast your heart is beating by measuring your pulse. Your pulse is taken on the inside of your wrist or at your neck. You can measure how fast your heart is beating by measuring the number of beats per minute.

A more accurate measure of how difficult it is for your heart to pump blood around your body is your blood pressure. A blood-pressure measurement records the pressure of the blood on the walls of the arteries. It is recorded as two numbers, for example 120/80. The higher number is the pressure when the heart has fully contracted. The lower number is when the heart is fully relaxed.

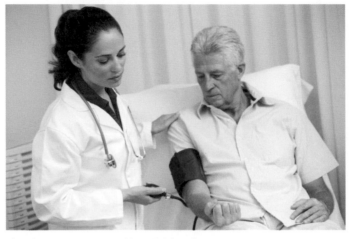

The doctor measures Oliver's blood pressure.

Narrowing of the arteries raises heart rate and blood pressure. Drugs such as caffeine, alcohol, cannabis, and Ecstasy have the same effect. High blood pressure increases the risk of CHD and heart attack.

What causes coronary heart disease?

It's usually easy for doctors to find the cause of a disease resulting from an infection. The pathogen that has infected the patient's body can be isolated and identified.

However, it's harder to find the cause of a non-communicable disease such as CHD. Non-communicable diseases are not caused by infections. There isn't one cause – but there are many **risk factors**. The more risk factors you are exposed to, the higher your risk of developing CHD.

Oliver has a family history of CHD. He is also overweight, smokes, and often eats high-fat, high-salt food. This diet has given Oliver high blood pressure and high cholesterol levels. Oliver does like sport – but he'd rather watch it on TV than do exercise himself. All these factors increase his risk of CHD.

Historical and global perspectives

One hundred years ago infections killed most people in the UK. Today better hygiene, vaccinations, and healthcare mean infections are more controlled. Non-communicable diseases are much more common than they were because of our modern lifestyles. In other parts of the world lifestyles are different, making other diseases more common.

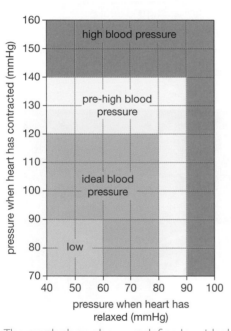

The graph shows how we define low, ideal, and high blood pressure. Many factors affect a person's blood pressure.

A diet high in saturated fat can cause high blood cholesterol – a risk factor for cardiovascular diseases.

How do we know about risk factors?

In 1948, a study of heart disease began in the town of Framingham, Massachusetts, USA. The study recruited 5209 men and women between the ages of 30 and 62. Since then, their children and grandchildren have also joined the study – over 13 000 people in total.

Every two years the researchers record details of each person, such as:

- body mass
- blood pressure
- cholesterol level
- lifestyle factors (e.g., whether they smoke and how much exercise they do).

Looking at the health of lots of people can show scientists the risk factors for different diseases.

The graph shows some data from the Framingham study. As cholesterol level increases, the death rate also increases. There is a **correlation** between the factor (cholesterol level) and the **outcome** (death rate). A claim that the factor is the **cause** of the outcome must be supported by a mechanism to explain the correlation. In this case, high cholesterol levels increase fatty deposits in the coronary arteries, which can lead to a deadly heart attack.

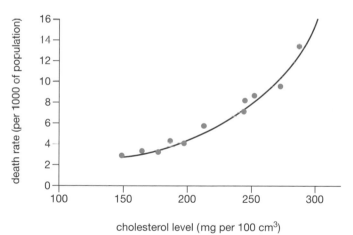

Data from the Framingham study.

The Framingham study has been very important for heart disease research. It has led to the identification of many major risk factors for heart disease.

Major risk factors for cardiovascular diseases

- smoking
- lack of exercise
- diet high in fat or salt
- being overweight or **obese**
- high blood cholesterol
- high blood pressure
- genes (family history of heart disease)

What makes a good study?

There are many reports in the media about studies of risk factors for diseases. You may want to use this information to make a decision about your own health, so it's important to know if a study has been done well.

Sampling

It's usually not possible for a health study to observe every person in the country. For this reason a study usually looks at a **sample** of people.

A good study looks at a large sample. This means the results are less likely to be affected by chance. For example, imagine a sample of five people with CHD. Most of the people have high blood pressure. Also, by chance, a few of them are called Bob. Is having high blood pressure, or being called Bob, linked to having CHD? We can't be sure from the data. A better sample size would be 1000 people with CHD. If most of them have high blood pressure, and a few are called Bob, it is clearer that the correlation between high blood pressure and CHD may be important.

Some studies focus on a particular group of people; for example, a study of blood pressure in middle-aged men. The study could look at a sample of 1000 middle-aged men. But the sample should not be 1000 middle-aged male football fans, because perhaps watching football raises blood pressure. This would be a biased sample. Equally, it would be biased to avoid football fans. Either way, the sample would not represent the population of middle-aged men.

Conclusions about a sample can only be applied to the whole population if it is a **representative sample**. This means that the types of people in the sample are similar to the types of people in the population. Choosing a sample at random can help.

Matched groups

Studies such as the one in Framingham follow the health of a particular group of people. A different type of study compares two groups of people. One group has the risk factor and the other doesn't, for example, a study comparing the health of people who smoke with people who don't. In these studies it's important to match the people in the two groups on as many other factors as possible. They should be the same age, sex, weight, and so on. Differences in outcomes between the two groups are then more likely to be due to the factor being investigated (e.g., smoking).

Can studying the health of these middle-aged, male football fans tell us anything about the health of the whole population?

Questions

1 List four lifestyle factors that increase a person's risk of CHD.

2 Explain why heart muscle cells need a good blood supply.

3 Explain how too much fat in a person's diet can lead to a heart attack.

4 Heart disease is more common in the UK than in developing countries. Suggest why.

5 Explain the difference between correlation and cause.

Key words

➤ sample
➤ representative sample

B: Lung cancer

Government health warnings have been printed on cigarette packets since 1971. There was evidence showing a correlation between smoking and lung cancer. In recent years the warnings have become stronger, smoking in public places has been banned, and shops have had to stop displaying tobacco products. How did scientists and the Government become convinced that smoking causes lung cancer?

Find out about
● risk factors for lung cancer
● developing explanations and peer review

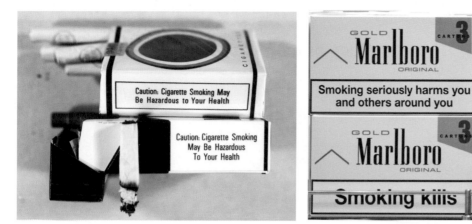

Health warning in 1971.

Modern health warnings.

An early clue

In 1948 a medical student in the USA, Ernst Wynder, observed the autopsy of a man who had died of lung cancer. He noticed that the man's lungs were blackened. There was no evidence that the man had been exposed to air pollution from his work. The man's wife said that he had smoked 40 cigarettes every day for 30 years.

So perhaps there was a link between smoking and lung cancer. But Wynder knew that one case is not enough to show a link between any two things.

In 1950 two British scientists, Richard Doll and Austin Bradford Hill, started a series of scientific studies. First, they compared people admitted to hospital with lung cancer to another group of people in hospital for other reasons. Smoking was very common at the time, so there were lots of smokers in both groups. But the percentage of smokers in the lung cancer group was much greater.

These results showed a correlation between smoking and lung cancer. Doll and Hill suggested smoking caused lung cancer. But a correlation doesn't always mean that one thing causes another.

How reliable was the claim?

Doll and Hill published their results in a medical journal so that other scientists could look at them. This is called **peer review**. Other scientists look at the data and how it was gathered. They look for faults and if they can't find any, then they are more likely to accept the claim.

The claim is also more likely to be accepted if other scientists can produce data that suggests the same conclusions. In this case, the data are said to be reproducible.

Key word
➤ peer review

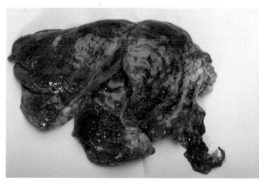

Lung tissue blackened by tar from cigarette smoke.

Number of cigarettes smoked per day	Number of cases of cancer per 100 000 men
0–5	15
6–10	40
11–15	65
16–20	145
21–25	160
26–30	300
31–35	360
36–40	415

The data show that there is a correlation between the number of cases of lung cancer in men and the number of cigarettes smoked.

Questions

1 Sketch a graph to show how the number of cases of lung cancer in men is affected by the number of cigarettes smoked.

2 Explain briefly what happens during peer review.

3 It's unlikely that many people would have agreed with Wynder if he had reported the case he saw in 1948. Suggest two reasons why.

4 If a man smokes 20 cigarettes a day from age 16 to 60, will he definitely develop lung cancer? Explain your answer.

A major study

In 1951 Doll and Hill started a much larger study. They followed the health of more than 40 000 British doctors for over 50 years.

The results were published in 2004 by Doll and another scientist, Richard Peto. They showed that:

● smokers die on average 10 years younger than non-smokers

● stopping smoking at any age reduces this risk.

The last piece of the puzzle – an explanation

Lung cancer rates in the USA rose sharply after 1930. The same pattern was seen in the UK.

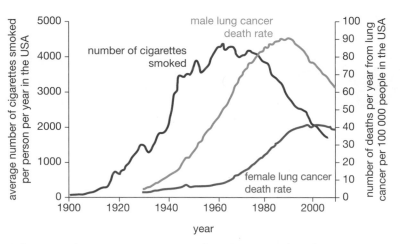

Before 1930 lung cancer was rare in the USA. As smoking became more popular with men, the number of lung cancer cases rose. This happened later for women because very few women smoked until after World War II.

Many doctors were now convinced that smoking caused lung cancer. But cigarette companies did not agree. They said other factors could have caused the increase in lung cancer, for example, more air pollution from motor vehicles.

The missing piece of the puzzle was an explanation of how smoking caused cancer. In 1998 scientists were able to explain how substances in cigarette smoke damage cells in the lungs, causing cancer. This confirmed that smoking causes cancer.

You may have heard about a person, perhaps a family member, who smoked all their life and never developed lung cancer. But one case is not convincing evidence against the correlation between smoking and lung cancer, just as Wynder's first case in 1948 was not enough evidence for it. Scientists do not usually give up an explanation if new data disagree with it, especially if the data come from just one case. An explanation usually survives until a better one has been developed.

C: Interacting factors

Coronary heart disease

In B2.4A you learnt that lifestyle is an important risk factor for CHD. A large-scale genetic study by the Wellcome Trust Case Control Consortium in 2007 compared the genomes of 2000 people with CHD to the genomes of 3000 people without. The researchers identified six genetic variants associated with CHD.

The risk that a person will develop CHD is affected by the genetic variants they inherited *and* their lifestyle. The interaction of these factors determines the overall risk. The more risk factors you are exposed to, the more likely it is you will get the disease. This is true for other non-communicable diseases as well. There is nothing you can do about your genome, but a person who has one or more 'risky' genetic variants can still reduce their risk of disease by adopting a healthy lifestyle.

Cancer

Asbestos is a mineral that was used in the construction of some old buildings.

Substances in cigarette smoke increase the risk of cancer. They do this by causing damage to the DNA in your cells. These cells become cancer cells. A substance that does this is a **carcinogen**.

> ### Synoptic link
>
> You can learn more about how cells become cancer cells in B4.3A *Cell division and growth.*

Research shows that if a person stops smoking, their risk of developing lung cancer decreases. After 15 years their risk can be similar to that of somebody who has never smoked. But it's not just cigarette smokers who are at risk. Breathing in other people's smoke and smoking cannabis are also dangerous.

There are carcinogens in our environment. People who worked with asbestos have a greater risk of lung cancer, and the risk is even higher if they smoked as well.

Your DNA accumulates damage as you get older, so age is a risk factor for cancer. The genetic variants you inherit, and exposure to **ionising radiation** can also increase your risk of various cancers. Ionising radiation from the sun increases the risk of skin cancer.

Find out about

- risk factors in the genome, the environment, and people's lifestyles
- how the interaction of these factors increases or decreases the risk of disease

Key words

➤ carcinogen
➤ ionising radiation

ANGELINA CANCER RISK

Actress Angelina Jolie has had surgery after being told that she had an 87% chance of developing breast cancer. She inherited a faulty gene from her mother, who died of ovarian cancer at the age of 56.
Jolie chose to have surgery to reduce her risk of breast and ovarian cancer. But a healthy diet, not smoking, and regular health checks can also reduce the risk of cancer.

Family history of a disease could mean you have inherited genetic variants that increase your risk. But there are ways to reduce your risk.

Key word

➤ body mass index (BMI)

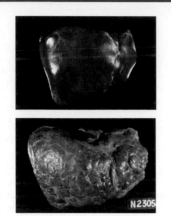

Healthy liver tissue (top), and liver tissue damaged by disease.

Liver disease

The liver is an organ that performs many important functions. It removes toxins from your body and helps to control blood cholesterol levels. Unfortunately, over 2 million people in the UK suffer from liver diseases.

There are different types of diseases that damage the liver, including:

- alcohol-related liver disease – caused by years of drinking too much alcohol
- fatty liver disease – caused by a build-up of fat within liver cells, usually in people who are overweight or obese
- inherited liver disease – a build-up of iron around the liver, caused by a faulty version of a gene called *HFE*.

Not drinking too much alcohol can help to keep the risk of liver disease low. It's also important to maintain a healthy **body mass index (BMI)**. Your BMI indicates whether your body mass is healthy for a person of your height.

Worked example: Calculating BMI

A 16-year-old boy is 171 cm tall and his mass is 68 kg. Does he have a healthy BMI?

Step 1: Write down what you know, with the units.

body mass = 68 kg

height = 171 cm = 1.71 m

Step 2: Write down the equation you will use to calculate the BMI.

$$BMI = \frac{body\ mass\ (kg)}{[height\ (m)]^2}$$

Step 3: Substitute the quantities into the equation and calculate the BMI.

$$BMI = \frac{68\ kg}{(1.71\ m)^2}$$

$$= \frac{68\ kg}{1.71\ m \times 1.71\ m}$$

$$= \frac{68\ kg}{2.9241\ m^2}$$

$$BMI = 23.2550186\ kg/m^2$$

Step 4: The body mass and height were given to the nearest whole number, so round the BMI to the nearest whole number.

$$BMI = 23\ kg/m^2$$

(to nearest whole number)

Work out whether this BMI is healthy by comparing it to a BMI chart. There are different charts for boys (aged 2–20), girls (aged 2–20), and adults. For a 16-year-old boy a healthy BMI is between about 16.1 and 24.2.

Answer:

The boy's BMI is 23 kg/m², which is in the healthy range for a boy of his age.

Questions

1 Explain how a person who has a 'risky' genetic variant can reduce their risk of CHD.

2 Write down four risk factors for cancer.

3 Roger is 22 years old. A genetic test has shown that he has a faulty version of a gene called *HFE*. He is very worried that he will develop liver disease.
   What advice would you give to Roger?

D: Interacting diseases

Different types of disease can interact. Having one disease can increase, or even decrease, the risk of developing or catching another. If having one disease leads to another, doctors call the second disease a **complication** of the first disease.

HPV and cervical cancer

Human papilloma virus (HPV) is the name of a group of viral pathogens. Some of the pathogens in the group cause cervical cancer in women. The cancer itself cannot be passed on to others, but HPV is a common STI.

Type 2 diabetes and cardiovascular disease

Sugar is absorbed from the food you eat and transported in the blood. When blood sugar levels are high, the pancreas releases a hormone called **insulin**. This causes sugar to be removed from the blood. People with type 2 diabetes cannot make enough insulin, or their body doesn't respond to it. This means their blood sugar level can rise to dangerously high levels. You will learn more about diabetes in B5.

Type 2 diabetes is not caused by a pathogen. People who are overweight or obese have a higher risk of developing the disease. At least 36 genetic variants have also been identified as risk factors. Type 2 diabetes itself increases the risk of developing other diseases, including cardiovascular diseases and blindness.

Sickle-cell trait and malaria

Sickle cell disease is an inherited disease that causes red blood cells to form a sickle shape. This damages their ability to carry oxygen around the body. The disease is common in many countries in Africa. It is caused by a recessive genetic variant. The disease occurs when a person inherits two copies of the variant (is homozygous).

However, if a person inherits one recessive variant and one dominant variant (is heterozygous) they have **sickle-cell trait**. This gives them some resistance to the pathogen that causes malaria. People with sickle-cell trait are less likely to die of malaria.

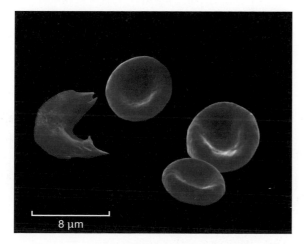

Three normal red blood cells and one that is sickle-shaped.

Find out about

- interactions between different types of disease

Key words

➤ complication
➤ insulin
➤ sickle-cell trait

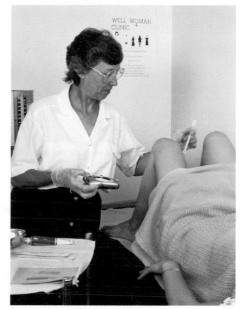

All women in the UK aged between 25 and 64 can have a cervical screening test (also known as a smear test) once every few years. It can detect abnormal cells and help prevent cervical cancer.

Questions

1 Explain why the UK vaccination programme includes a vaccination against HPV for young girls.

2 Describe an example of an interaction between two non-communicable diseases.

3 Malaria is widespread in the African country of Uganda. A child inherits one recessive variant of the sickle cell gene. Explain how this could be an advantage in Uganda.

A: The best medicine

Find out about

- the use of medicines in the treatment of disease
- why antibiotics are becoming less effective
- lowering the risk of coronary heart disease
- treating coronary heart disease

Key words

➤ medicine
➤ antibiotics
➤ antibiotic resistance

Jake finds out that the doctor won't give him any antibiotics.

Antibiotics are made by some bacteria and fungi to destroy other microorganisms. The fungus growing on this bread makes penicillin.

Remember the doctor's waiting room from the beginning of this chapter? Some of the people in the waiting room may need a **medicine** to help them feel better. The human race has been using medicines to treat diseases for thousands of years.

We have discovered medicines from natural sources, such as plants and fungi. We have created medicines using chemical reactions. And we have modified the genomes of certain organisms so that they make medicines for us.

What does it do?

Some medicines treat the symptoms of a disease but not the cause. They can reduce the severity of the symptoms, so that we don't feel so ill. They can also reduce the length of time we experience symptoms, which makes us feel better sooner. Painkillers are an example of this kind of medicine. For a minor illness caused by an infection, this can be enough – the medicine treats the symptoms, while the body's white blood cells eliminate the cause (the pathogen).

Some medicines help to treat the cause of a disease. For example, some medicines kill pathogens or stop them reproducing. **Antibiotics** kill or inhibit bacteria – but they do not work against fungi or viruses.

Antibiotics

Jake from the doctor's waiting room has the symptoms of a cold. The doctor tells him that he won't be given any antibiotics. His cold is caused by a virus. Jake's own body will fight the viral infection by itself.

The Ancient Egyptians may have been the first people to use antibiotics. They put mouldy bread onto infected wounds. Scientists now know that the mould is a fungus that makes penicillin. In the 1940s, scientists started to grow the fungus to make larger amounts of penicillin. At first, penicillin was called a 'wonder drug'. Before the 1940s, bacterial infections had killed millions of people every year. Now these infections could be cured using antibiotics.

But within 10 years, one type of bacteria was no longer killed by penicillin. The bacteria had become resistant. The problem of **antibiotic resistance** has not gone away. Bacteria have developed resistance to many of the new antibiotics that have been discovered.

How do bacteria become resistant to antibiotics?

One change (mutation) in a single gene can be enough to make a bacterium resistant to an antibiotic. One resistant bacterium on its own won't do much damage. But if it reproduces rapidly, it can produce a large population of bacteria all resistant to an antibiotic. Some fungi, such as those that cause ringworm and thrush, have become resistant to antifungal substances in the same way.

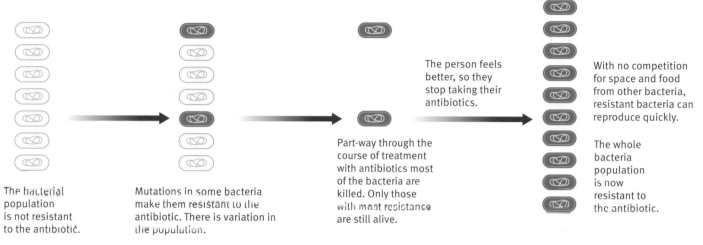

The bacterial population is not resistant to the antibiotic.

Mutations in some bacteria make them resistant to the antibiotic. There is variation in the population.

Part-way through the course of treatment with antibiotics most of the bacteria are killed. Only those with most resistance are still alive.

The person feels better, so they stop taking their antibiotics.

With no competition for space and food from other bacteria, resistant bacteria can reproduce quickly.

The whole bacteria population is now resistant to the antibiotic.

Genetic mutations can result in antibiotic-resistant bacteria.

Rise of the 'superbugs'

Some bacteria, such as MRSA, are resistant to multiple antibiotics. They have become known as 'superbugs'. Some superbugs may soon become resistant to all known antibiotics. They pose a very serious danger to your health.

Superbugs are becoming increasingly common in hospitals, where antibiotics are used every day. But resistance can appear anywhere antibiotics are used.

Two things increase the likelihood of antibiotic resistance developing:

- people taking antibiotics they don't really need
- people not finishing their course of antibiotics.

Scientists cannot stop antibiotic-resistant bacteria from developing. The mutations that produce superbugs are part of a natural process. We must hope that scientists can develop new antibiotics fast enough to keep us one step ahead of the bacteria.

But there are things we can do to slow down the development of antibiotic resistance:

- have better hygiene in hospitals to reduce the spread of resistant bacteria
- only prescribe antibiotics when a person really needs them
- always finish a course of antibiotics (unless harmful side-effects develop).

If you are given a course of antibiotics and take them all, it is likely that all the harmful bacteria will be killed – including any resistant ones.

'TICKING TIME BOMB'

England's chief medical officer has called antibiotic resistance a "ticking time bomb". The Prime Minister warned the world could be "cast back into the dark ages of medicine".
Without effective antibiotics, even minor surgery and routine operations could become high-risk procedures.

SUPERBUGS ON THE RAMPAGE

A government report has warned that an outbreak of an antibiotic-resistant infection could affect 200 000 people. Two in every five infected people could die. The report says the number of infections will "increase markedly over the next 20 years".

Antibiotic resistance is headline news.

Is it resistant?

Bacteria can be tested in the laboratory to see whether they are resistant to an antibiotic. The bacteria are grown in a Petri dish containing nutrient agar jelly. A disc of antibiotic is placed on the agar. Molecules of antibiotic diffuse outwards from the disc. After several days, a clear zone may be seen around the disc. This zone appears where the antibiotic has killed the bacteria. However, if no clear zone appears you can conclude that the bacteria are resistant to the antibiotic.

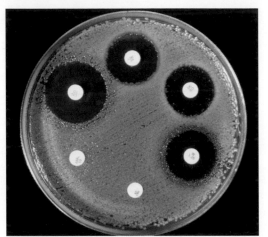

This Petri dish contains bacteria grown on agar jelly. Each white disc contains a different antibiotic. Two of the antibiotics have not killed the bacteria.

Worked example: The area of a clear zone

One way to compare the effectiveness of different antibiotics is to compare the sizes of the clear zones they produce. We can do this by working out the area of each zone.

Step 1: Measure the diameter of a clear zone. The diameter is the distance across the clear zone from one edge to the other (including the antibiotic disc).

diameter

diameter = 20 mm

Step 2: Work out the radius (*r*) of the clear zone. You will need the radius to work out the area. The radius is half the diameter.

$$r = \frac{\text{diameter (mm)}}{2}$$

$$= \frac{20 \text{ mm}}{2}$$

$$r = 10 \text{ mm}$$

Step 3: Write down the equation you will use to calculate the area.

$$\text{area} = \pi \times r^2$$

Step 4: Substitute the quantities into the equation and calculate the area. Your calculator might have a π button. If it doesn't, you can use the value 3.14 for π.

$$\text{area} = 3.14 \times (10 \text{ mm})^2$$

$$= 3.14 \times 100 \text{ mm}^2$$

$$\text{area} = 314 \text{ mm}^2$$

Step 5: Repeat steps 1–4 for each of the other clear zones, then compare the areas.

Answer:

The clear zone with the biggest area is the most effective antibiotic.

Treating cardiovascular disease

In B2.5 you learnt about CHD, which is a type of cardiovascular disease. There is no cure for CHD. However, there are things that you can do both before and after diagnosis to help keep your heart and blood vessels healthy.

Prevention is better than treatment!

As with any disease, it's best to stay healthy and avoid getting CHD in the first place. Eating a healthy diet (low in fat and salt), taking regular exercise, and not smoking will all help lower your risk of CHD. The sooner you make these changes to your lifestyle, the better.

Treatment options

For a person who has been diagnosed with CHD, it is not too late to make lifestyle changes. These will still reduce the risk of heart attack, and they can help to lower blood pressure. People with very high blood pressure may also be prescribed medicines to help lower it. The patient may have to take the medicine for the rest of their life. However, some people can experience harmful reactions to medicines.

If the coronary arteries have become blocked, surgery may be necessary. If the heart has been severely damaged, the only option may be a heart transplant.

A doctor will have to evaluate the treatment options to decide which one is best for each patient. Many factors need to be considered, including:

- the likely effectiveness of the treatment
- the risk of causing further harm
- the costs and benefits to the patient and the health service.

Questions

1 Explain why Jake's doctor will not give him any antibiotics to treat his cold.

2 Write bullet points to explain how failure to complete a course of antibiotics can lead to the spread of antibiotic resistant bacteria.

3 Describe two things that you can do to reduce the risk of antibiotic-resistant bacteria developing.

4 Describe and explain how bypass surgery reduces the risk of heart attack for a person with CHD.

5 Make a table summarising the benefits, risks, and costs of the following treatment options for CHD:
   - lifestyle changes
   - medicines
   - surgery.

   For each treatment option, identify the benefits, any risks, and the costs to the patient and to the health service.

HEALTHY HEART

- Cut down on fatty foods to lower blood cholesterol.
- If you smoke, stop.
- Maintain a healthy BMI to avoid high blood pressure and reduce strain on the heart.
- Take regular exercise (such as 20 minutes of brisk walking each day) to increase the fitness of the heart.
- Reduce the amount of salt eaten to help lower blood pressure.
- If necessary, take medicines to reduce blood pressure and/or cholesterol level.
- Relax, reduce stress.

The best ways to keep your heart and blood vessels healthy.

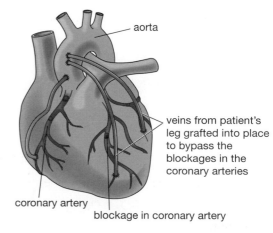

Heart surgery is complicated; there are risks of bleeding, infection, damage to tissues and organs, stroke, and death.

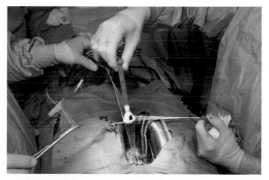

aorta

veins from patient's leg grafted into place to bypass the blockages in the coronary arteries

coronary artery

blockage in coronary artery

Bypass surgery can help reduce the risk of heart attack from blocked coronary arteries.

B: Developing new medicines

Find out about

- the discovery and development of potential new medicines
- pre-clinical and clinical testing of medicines

Key words

➤ target
➤ screening
➤ pre-clinical testing
➤ clinical testing
➤ control group

Most of us take medicines prescribed by our doctor without thinking much about it. But what if we could ask the scientist who developed the medicine some questions?

Understanding how new medicines are developed can help us answer these questions.

The need for new medicines

Many diseases do not currently have treatments. Some medicines do not work as well as we would like, and some medicines cause harmful side-effects. Pathogens change, making existing medicines – such as antibiotics – less effective. So there is an ongoing, and sometimes urgent, need to develop new medicines.

Drug discovery

Nature is a valuable source of new medicines. But we cannot rely on scientists climbing through jungles with collecting pots – we would never discover new medicines quickly enough.

New antibiotics have been found in surprising places, including in crocodile blood.

Studying the genomes and proteins of a pathogen and the organism it infects can suggest a **target** for a new medicine. The target could be an antigen or receptor molecule on the pathogen or a host cell. The process of drug discovery usually begins by testing large libraries of substances to find out which ones react with the target. This is called **screening** and can be done quickly by a machine. It is unlikely that a perfect medicine will be found, but promising substances are selected for modification and further screening.

Eventually, some of the substances may be chosen to be made into new medicines. They enter the process of drug development.

Sian is researching possible new treatments for cancer.

Drug development: pre-clinical testing

All new medicines have to be tested on living organisms before they are made widely available. The first stage of this process is **pre-clinical testing**.

A new medicine is first tested on human cells cultured (grown) in a laboratory. Scientists try out different concentrations of the medicine. They test it on different types of body cells with the disease. These tests check how well the medicine works against the disease – how *effective* it is. They also give the scientists data about how *safe* the medicine is for the cells.

If the drug passes tests on human cells, it is tested on animals. These tests make sure the drug works as well in whole animals as it does on cultured cells.

New medicines are first tested on cells grown in a laboratory.

Pre-clinical testing includes tests on animals.

Drug development: clinical testing

If a new medicine passes pre-clinical testing it can then be tested on people. This is called **clinical testing** or clinical trials.

Scientists first test the medicine on healthy human volunteers. This shows how *safe* the medicine is. They then test it on people with the disease. This shows how *safe* the medicine is, and how *effective* it is.

Long-term human trials check that the medicine is effective and that it does not cause harmful side-effects when used for a long time. These studies may continue after a drug is approved for use.

Not everybody agrees that it is right to test medicines on animals. The British Medical Association (BMA) believes that animal experiments are necessary at present to develop a better understanding of diseases and how to treat them. However, it says alternative methods should be used whenever possible.

Treatment and control groups

People who agree to take part in a clinical trial will be put randomly into one of two groups. One group will be given the new medicine – this is the treatment group. The other group will not get the new medicine – this is the **control group**. In almost all clinical trials the control group is given the treatment that is currently being used.

At the end of the trial, the results from the two groups are compared.

Comparing the results from the groups shows whether the new treatment is any better than the current one.

I know the trial could help people in the future – but what about me? Could I be risking my health? Will I be given the new medicine or not?

Anna has questions about taking part in a clinical trial.

Key words
➤ placebo
➤ double-blind trial
➤ blind trial
➤ open-label trial

For a person with a disease, taking the new medicine could help them get better, but it may bring other risks. If a trial shows that the risks are too great, it is stopped.

Sometimes there isn't any current treatment for a disease. In these cases the control group can be given a **placebo**. This looks exactly like the new medicine but has no medicine in it.

Taking a placebo will not help a person with the disease to recover. While taking the placebo they will miss out on any benefits of the new medicine. Is it ethical to give a placebo to a person with a disease? Clinical trials that give a placebo to people with a disease are rare. When a trial shows that a new medicine has benefits it is immediately offered to the control group.

Who knows what?

In some clinical trials, neither the doctor nor the patient knows whether the patient is receiving the new medicine or a placebo. If the doctor or patient did know, it could affect the way they report the patient's symptoms. Someone else prepares the treatments. This is a **double-blind trial**.

In some trials the doctor is told which patients are being given the new medicine. This may be because the doctor needs to look very carefully for certain harmful side-effects. The patient is not told what they are given. This is a **blind trial**.

In other trials, both the patient and the doctor know who receives the new medicine. This is an **open-label trial**. This may be necessary if there is no way to hide which treatment is the new medicine.

An open-label trial may also be used when the new medicine is given to *all* the patients in the trial. This happens when there is no other treatment and patients are so ill that doctors are sure they will not recover from the illness. In this case it would be wrong not to offer the hope of the new drug to all the patients.

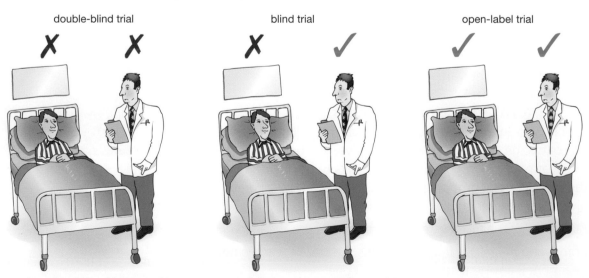

double-blind trial   blind trial   open-label trial

In a clinical trial the doctor and/or patient may (✓) or may not (✗) know if the treatment is the new medicine.

The final product

If a new medicine is found to be safe and effective during clinical trials, the manufacturer will want to sell it. In the UK they must apply for authorisation from the Medicines and Healthcare products Regulatory Agency (MHRA). They will also apply for a patent – this stops anybody else making and selling the medicine for 20 years.

The discovery and development process can take 10 years. From thousands of substances, perhaps only one will become a medicine. It is a very expensive process. Selling medicines, rather than giving them freely to everybody who needs them, pays for the discovery and development of new medicines.

Modelling the effects of medicines

It would be dangerous to put a new medicine straight into a human. So it is tested on cells and animals first. They enable us to model what could happen in a human.

Some scientists also use mathematical models. Living organisms are complex systems of interactions between molecules. The medicine may cause a change in the amount of molecule X, which in turn affects the amount of molecule Y, and so on. We can model these changes mathematically using a computer.

Cell, animal, and computer models have some, but not all, of the characteristics of a human. We use them to collect data on the safety and effectiveness of a medicine. They enable us to investigate medicines more quickly, more cheaply, and with fewer ethical questions than using humans. But their usefulness is limited because they do not represent a human exactly.

Drug Discovery

studying genomes and proteins of pathogens and host cells to identify target

5000 to 10 000 substances

screening to find substances that react best with the target

approximately 250 substances

Drug Development

pre-clinical testing on human cells for safety and effectiveness

pre-clinical testing on animals for safety and effectiveness

up to 10 substances

clinical testing on healthy human volunteers for safety

clinical testing on humans with the disease for safety and effectiveness

one or two substances

MHRA authorisation and patent protection

one substance

medicine that can be sold and prescribed to patients

An overview of the drug discovery and development process.

Questions

1 Copy and complete the table.

Stage	Medicine tested on:	To find out:
	human cells grown in a laboratory	• how safe the drug is for human cells • how well it works against the disease
	animals	
clinical testing		• how safe the drug is for humans
	humans with the disease	

2 Explain the difference between blind and double-blind trials.

3 Describe a situation in which it would be ethically wrong to use a placebo in a trial.

4 A new medicine is compared to a physiotherapy treatment. What kind of trial would be used?

5 Explain why it is beneficial to protect new medicines with patents.

6 Anna has Huntington's disease. She is unsure about taking part in a trial of a new medicine for the disease. What should she consider when making her decision?

Science explanations

B2 Keeping healthy

In this chapter you learnt about the relationship between health and disease. Keeping healthy depends upon avoiding infection, the defences in your body, maintaining a healthy lifestyle, and using medicine when necessary.

You should know:

- that communicable diseases are caused by infection with pathogens (bacteria, viruses, protists, and fungi), and they can be spread from organism to organism
- for at least one common human infection, one plant disease, and HIV/AIDS:
  - which type of pathogen causes the infection
  - how the pathogen is spread
- that the skin and mucus, stomach acid, saliva, tears, and helpful bacteria are non-specific defences of the human body against pathogens
- how white blood cells protect us against pathogens, including:
  - ingesting and digesting them
  - making antibodies to disable pathogens or tag them for attack
  - staying in the body as memory cells to make us immune to the pathogen in the future
- how the spread of diseases can be reduced and prevented in humans and plants, including sexually transmitted infections
- how vaccination helps to prevent disease
- that non-communicable diseases are not caused by infections, but a person's risk of developing these diseases is affected by their genome, their environment, and their lifestyle
- about risk factors for cardiovascular diseases, cancer (including lung cancer), liver disease, and type 2 diabetes
- how some different diseases can interact
- about ways to treat disease, including the use of medicines such as antibiotics
- about the problem of bacterial resistance to antibiotics and what we can do about it
- how new medicines are discovered, developed, and tested.

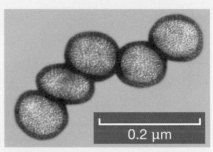

Flu is caused by infection with the influenza virus.

Ideas about Science

Stories about health and disease are always in the news. When deciding which health reports to trust, it is useful to know about correlation and cause. A correlation is a link between a factor and an outcome. For example, there is a correlation when:

● an outcome happens when a factor is present, but not when it is absent

● an outcome increases or decreases when a factor increases.

When a correlation has been observed, the factor may be the cause (or one of the causes) of the outcome. In many situations, the factor may not lead directly to the outcome but will increase the chance (or the risk) of it happening. For example, there is a correlation between smoking cigarettes and developing lung cancer. People who smoke will not definitely develop lung cancer, but their risk of developing it is much greater.

In order to claim that a factor causes an outcome, we need to identify a process or mechanism that explains the link. In this case of smoking cigarettes and lung cancer, substances in the cigarette smoke damage cells in the lungs, which causes the cancer.

You should be able to:

● identify a correlation in data

● suggest factors that increase the chance of an outcome, but do not always lead to it

● explain why one case is not enough evidence for or against a correlation

● explain why you might reject a claim that a factor causes an outcome (usually because there is no mechanism to link the factor and the outcome).

Findings reported by a scientist or group are checked by the scientific community before being accepted. This is peer review. Claims about health and disease are stronger if they have been peer reviewed. Scientists are usually sceptical about claims based on results that cannot be reproduced by anyone else, and about unexpected findings, until they have been repeated (by themselves) or reproduced (by someone else).

There are ethical questions about testing new medicines, for example, when using animals and when using placebos. You should be able to:

● state clearly what the ethical issues are

● discuss different views that may be held.

Vaccination is the best choice for society as a whole. But individuals have the right to decide whether they get vaccinated.

B2 Review questions

1 Look at the list of diseases.

- influenza
- cardiovascular disease
- HIV/AIDS
- *Salmonella* food poisoning

   a Which of the diseases are communicable?

   b Which of the diseases is a sexually transmitted disease?

   c Which of the diseases could be treated using antibiotics?

2 Our bodies have defences that make it difficult for pathogens to enter the blood.

   a Describe two physical defences of the human body against pathogens.

   b Explain why these defences are called 'non-specific defences'.

3 Explain why catching a cold does not make you immune to catching another cold in the future.

4 Describe how the spread of communicable diseases can be reduced or prevented in plants.

5 We can protect ourselves against infection by having a vaccination. The vaccine contains dead pathogens.

   Some of the statements below describe how a vaccination makes us immune to the pathogen. Not all the statements are correct.

   A The pathogens are destroyed.

   B White blood cells make lots of antigens.

   C The dead pathogens turn into memory cells.

   D Memory cells stay in the blood so they can react quickly in case of re-infection.

   E The dead pathogens enter the blood.

   F The dead pathogens release antibodies into the blood.

   G White blood cells make antibodies against the pathogen's antigens.

   Select the four correct statements and then write their letters in the correct order.

6 Eating a diet containing a lot of fatty food can increase the risk of heart disease. The four people shown in the margin have different views about this.

To answer these questions, you may use each person once, more than once, or not at all.

a Which person says that the absence of replication is a reason for questioning a scientific claim?

b Which person is suggesting that individual cases do not provide convincing evidence for or against a correlation?

c Which person is describing the process of peer review?

d Which two people are suggesting that factors might increase the chance of an outcome but not always lead to it?

Jane
Eating fatty foods puts you at higher risk of getting heart disease. But it doesn't make it certain to happen.

Ranjit
My grandad ate fatty food all his life. He died of influenza at 83. Scientists look at lots of data before they conclude that a high-fat diet increases the risk of heart disease.

Peter
We only know that fatty foods can cause heart disease because many scientists have collected data. If there was only one study we would be less sure.

Stella
I am a food scientist. My findings are always checked by other scientists before they are published.

7 The bar chart shows the death rate from CHD in males and females of different races.

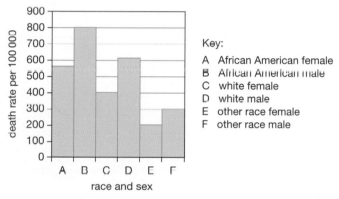

Key:

A African American female
B African American male
C white female
D white male
E other race female
F other race male

a Which race shown on the bar chart consistently has a higher death rate from heart disease?

b Race and sex are both risk factors for CHD. Write down four other factors that can affect a person's risk of developing CHD.

c The bar chart provides data on heart disease for African Americans. These are people of African origin who live in America.

Africans who still live in Africa show a much lower incidence of heart disease. Suggest why.

8 Outline how you would investigate the effectiveness of six different antibiotics against bacteria. Your answer should include:

a the equipment and materials you would use

b which factors you would control

c what measurements you would take

d how you would process the data.

9 New medicines are tested for effectiveness and safety. Explain what and who the medicines are tested on, and the role played by open-label, blind, and double-blind trials.

10 Doctors have to make decisions about when to use placebos and who they should be used on. Explain what placebos are, why they are used, and when they should not be used.

B3 Living together – food and ecosystems

Why study food and ecosystems?

Scientists estimate that there are approximately 8.7 million species of living organisms on Earth. Each species, and each individual, relies upon others for its survival. Almost every living organism depends on the ability of some organisms to make food by photosynthesis. Individuals and populations compete for resources, and rely on microorganisms to recycle elements vital to life. But human activities are changing ecosystems. We are all connected, and this means the impacts of environmental changes could endanger us all.

What you already know

- There are similarities and differences between plant and animal cells.
- Some organisms make their own food by photosynthesis.
- The inputs of photosynthesis are light, water, and carbon dioxide. The outputs are glucose and oxygen.
- Leaves are adapted for photosynthesis and stomata are important in gas exchange.
- Water and minerals enter a plant through the roots.
- Molecules of a dissolved substance move through the solvent by diffusion.
- There are differences between carnivores, herbivores, and omnivores, and between producers and consumers.
- Individuals of the same type living in the same place make up a population, and all the interacting populations in an ecosystem make up the community.
- Food chains and food webs show feeding relationships within a community.
- The organisms in a community depend on one another for food, and for the breakdown and recycling of substances.
- Animals can act as pollinators and seed dispersers for plants.
- Changes in an ecosystem can affect the survival of individuals and populations.

The Science

There's more to photosynthesis than its inputs and outputs. It's a series of chemical reactions involving enzymes, and it depends on the movement of molecules into and out of cells. The glucose made by photosynthesis is at the start of almost every food chain on Earth.

The interdependence and competition between organisms is critical to their survival, which can be affected – both negatively and positively – by changes within ecosystems.

Ideas about Science

We can use models to help explain complex interactions, such as those between enzymes and reacting molecules, and between predators and prey within a community.

When thinking about changes within ecosystems, we must understand how human activities can have unintended impacts on the interacting populations of organisms that live there.

A: Making food

Find out about

- how models help us explain what happens in photosynthesis
- how chloroplasts are related to photosynthesis
- why photosynthesis can be described as an endothermic process

Key words

➤ producer
➤ photosynthesis
➤ biomass
➤ chloroplast
➤ chlorophyll

Life on Earth depends on plants. They are producers of biomass, and release oxygen into the atmosphere.

A plant in a garden or a tree in a field may not look like it's doing much, but every living plant is a busy biological factory. Plants use chemical reactions to make substances vital to life on Earth. Without these reactions inside plant cells, humans and other animals would not exist.

What happens during photosynthesis?

As you know, a plant is a **producer**. It makes its own food in the form of glucose. The glucose is made from carbon dioxide and water by **photosynthesis**. This process needs light, which usually comes from the Sun. The glucose made in plants is the source of **biomass** for most ecosystems.

Oxygen is a by-product of photosynthesis. Plants use some of the oxygen in cellular respiration and the rest is released, as oxygen gas, into the atmosphere. It may be odd to think of oxygen as a waste product. Humans and most living organisms on Earth depend on it for aerobic respiration.

Photosynthesising organisms changed the atmosphere of the early Earth. They added oxygen to the atmosphere. The oxygen-rich atmosphere affected the evolution of life on Earth.

Models of photosynthesis

Describing photosynthesis as we just have is useful. It's a descriptive model that helps us understand the process. Other models of photosynthesis are also useful.

Models of photosynthesis: equations

One model of photosynthesis is a word equation or word summary. It uses words to represent the reactants and products.

$$\text{carbon dioxide} + \text{water} \longrightarrow \text{glucose} + \text{oxygen}$$

Another model of photosynthesis is a balanced chemical equation. It uses chemical formulae to represent the reactants and products.

$$6CO_2(g) + 6H_2O(l) \longrightarrow C_6H_{12}O_6(aq) + 6O_2(g)$$

Sometimes 'light' or 'energy' is written above the arrow, to show that the process needs energy from the Sun (or another light source).

These equations are convenient ways of describing the process of photosynthesis. However, they miss out much of the detail. They suggest that only one chemical reaction is involved. In fact, photosynthesis involves a number of chemical reactions. The equations do not tell us where the reactions of photosynthesis take place.

Many steps are involved in the process of photosynthesis. It is not just one chemical reaction. The steps do not all happen at the same time. A more detailed model can help us to better understand what happens.

Models of photosynthesis: a two-stage model

Photosynthesis can be summarised in two main stages.

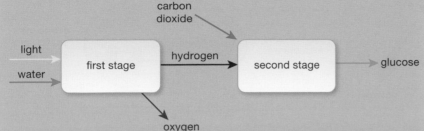

The first stage needs light. In this stage, water molecules are split into hydrogen and oxygen. The oxygen is released from the plant into the air.

The second stage does not need light. In this stage, the hydrogen is combined with carbon dioxide from the air to make glucose.

This two-stage model is a better representation of photosynthesis than a simple equation. But it still misses out a lot of detail about the process.

Inside a leaf cell

Leaf cells contain organelles called **chloroplasts**. Each chloroplast contains **chlorophyll**, a pigment that absorbs light and splits water into hydrogen and oxygen in the first stage of photosynthesis.

Chloroplasts also contain enzymes that are needed for photosynthesis. You'll learn more about enzymes in B3.1B. Some of the glucose made by photosynthesis is converted into starch for storage. It is stored in starch grains in chloroplasts.

Synoptic link

You can learn more about the effects of photosynthesis on the early atmosphere of the Earth in C1.1C *How did our atmosphere form?*

Given enough raw materials, light, and the right temperature, a large tree could produce 90 kg of glucose in a day.

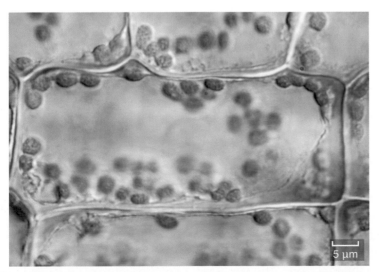

Cells from *Elodea* pondweed, seen under a light microscope. The dark green organelles are chloroplasts. Chloroplasts are approximately 5 μm in diameter.

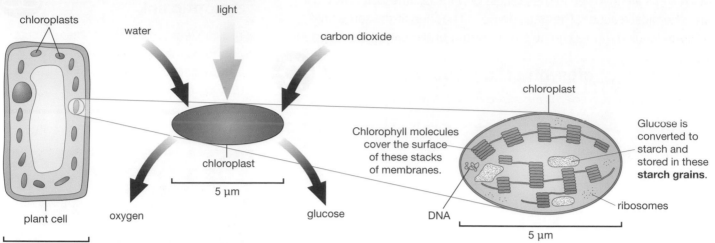

How chloroplasts in plant cells are related to photosynthesis.

Photosynthesis and energy

Photosynthesis only happens in the light. Energy from the Sun (or another light source) is needed to make the process happen.

One way of keeping track of energy is to think of **energy stores**. Energy in a store is a quantity in joules that you can calculate.

Synoptic link

You can learn more about energy stores in P2.1A *What's the big energy picture?*

There are several different energy stores involved in photosynthesis:

nuclear store	the Sun
chemical store	molecules of glucose plus oxygen

You can think of the Sun, a leaf, and its surroundings as a **system**. Photosynthesis is a change in the system. It needs energy from the Sun (and molecules of carbon dioxide and water) to make molecules of glucose and oxygen. Before the change, there is lots of energy in the nuclear store (the Sun). After the change:

- the energy in the nuclear store (the Sun) has decreased
- the energy in the chemical store (glucose and oxygen) has increased.

An endothermic process

Energy from the Sun warms things up. It increases the amount of energy in thermal stores. A dead leaf would be warmed by energy from the Sun in this way. But if the leaf was alive, it would not warm up so much. This is because of photosynthesis – some of the energy from the Sun would top up a chemical store (glucose plus oxygen); so less of the energy would top up thermal stores. The temperature of the leaf tissue is lower because of the reactions of photosynthesis. Because of this, photosynthesis can be described as an **endothermic** process.

During photosynthesis, energy from the Sun is used to top up chemical stores in leaves.

Key words

- ➤ energy store
- ➤ system
- ➤ endothermic

Questions

1 What is the waste product of photosynthesis?

2 **a** What is the name of the pigment needed for photosynthesis?
  b In which organelles is this pigment found?

3 Without using the box showing the two-stage model of photosynthesis, draw your own diagram to summarise the two-stage model.

B: Enzymes

You may have heard of 'biological' washing detergents. They contain an **enzyme**. These detergents are better at removing stains from clothes washed at lower temperatures.

Enzymes are found in the cells of all living organisms. Many chemical reactions, including photosynthesis, rely on enzymes.

What are enzymes?

An enzyme is a biological **catalyst** that speeds up a chemical reaction. Enzymes catalyse reactions that break down large molecules into smaller ones, and help to join small molecules together to make larger ones.

Enzymes are proteins, made up of long chains of amino acids. The sequence of amino acids in each enzyme is determined by instructions in a gene. The amino acid chains in different enzymes are folded up into different shapes. An enzyme's shape is very important to how it works.

How do enzymes work?

We can use a model to explain how enzymes work. It is called the **lock-and-key model**.

How enzymes work: the lock-and-key model

The molecule that is changed by an enzyme is called the **substrate**. The substrate molecule must be the correct shape to fit into a part of the enzyme called the **active site**. It's a bit like fitting the right key in the right lock. Because of this, enzymes are specific – they usually only work on one substrate.

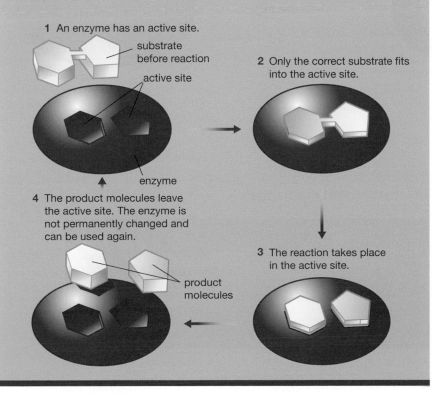

1 An enzyme has an active site.
   substrate before reaction
   active site
   enzyme

2 Only the correct substrate fits into the active site.

3 The reaction takes place in the active site.

4 The product molecules leave the active site. The enzyme is not permanently changed and can be used again.
   product molecules

Find out about

- the role of enzymes in photosynthesis and other processes
- how enzymes work
- factors affecting enzyme activity

Key words

➤ enzyme
➤ catalyst
➤ lock-and-key model
➤ substrate
➤ active site

A biological washing powder contains enzymes. They help to break down the substances in stains.

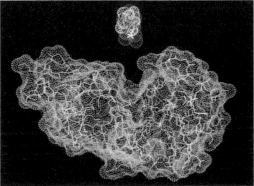

A computer model of an enzyme (bottom) and its substrate (top). In reality, the enzyme molecule is around 10 nm long. The enzyme molecule and the substrate molecule are the correct shapes to fit together – like a key in a lock.

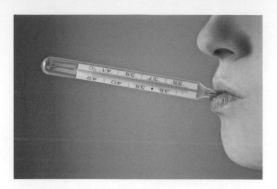

Enzymes in humans work best at around 37 °C. However, other organisms have cells and enzymes adapted to different temperatures, to suit the temperature of the environment in which they live.

Why do living organisms need enzymes?

At biological temperatures (from about 0 °C to 50 °C) many chemical reactions would happen too slowly to support life. But living organisms have managed to overcome this problem.

One way of speeding up a reaction is to increase the temperature. But higher temperatures damage living cells.

So, living organisms use enzymes to increase the **rate** of reactions in their cells. Enzymes can increase rates of reaction by up to 10 billion times. It is not possible to live without enzymes.

How and why does temperature affect enzyme reaction rates?

The rate of chemical reactions usually increases when the temperature increases. Enzyme reactions get faster if the temperature is increased from a low value. The temperature at which an enzyme works best is called its **optimum** temperature.

However, above the optimum temperature the reaction slows down. This is because enzymes are proteins. As the temperature rises, the enzyme's shape changes and it no longer works as well. We can explain what happens using the lock-and-key model.

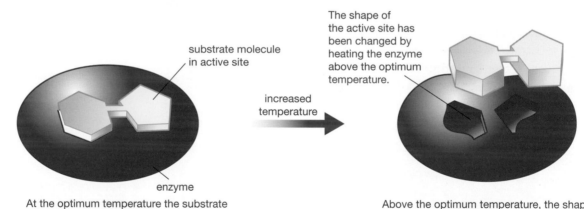

substrate molecule in active site

The shape of the active site has been changed by heating the enzyme above the optimum temperature.

increased temperature

enzyme

At the optimum temperature the substrate fits into the enzyme's active site.

Above the optimum temperature, the shape of the enzyme changes. The substrate no longer fits into the active site.

Very high temperatures permanently change the shape of an enzyme. Even when the enzyme cools down, it does not go back to its original shape. Like cooked egg white, the protein cannot be changed back. The enzyme is said to be **denatured**.

Microorganisms living in hot springs have enzymes that can work at temperatures up to around 120 °C.

Antarctic icefish are active at temperatures around 0 °C. Their enzymes are adapted to work best at these temperatures. Most organisms would be dead or very sluggish at 0 °C.

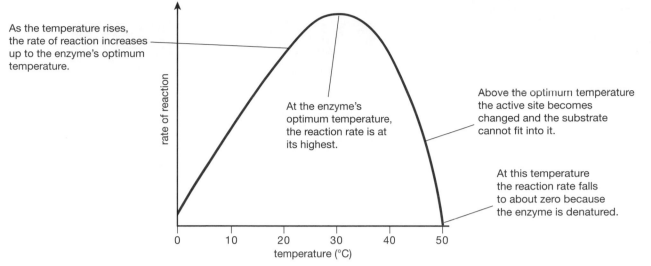

As the temperature rises, the rate of reaction increases up to the enzyme's optimum temperature.

At the enzyme's optimum temperature, the reaction rate is at its highest.

Above the optimum temperature the active site becomes changed and the substrate cannot fit into it.

At this temperature the reaction rate falls to about zero because the enzyme is denatured.

This sketch graph shows the rate of a reaction at different temperatures, for a reaction catalysed by a human enzyme. All other conditions were kept constant.

How and why does pH affect enzyme reaction rates?

The shape of proteins, including enzymes, can be affected by acids and alkalis. The shape of an enzyme's active site will be changed if bonds holding the protein chains together are broken. Many of these bonds depend on how acidic or alkaline the enzyme's environment is. In other words, they depend on the pH. When the shape of the enzyme changes, the substrate no longer fits in the active site. Every enzyme has an optimum pH at which it works best.

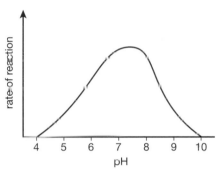

The enzyme used in this reaction has an optimum pH of 7.4.

How and why does substrate concentration affect enzyme reaction rates?

An enzyme works by joining with its substrate at its active site. If there is no substrate nothing will join and the reaction rate will be zero. With a low concentration of substrate, there will be the occasional coming together of substrate and enzyme, and the reaction rate will be slow. As more and more substrate is added, the reaction gets faster and faster. However, a point is reached when all the enzyme molecules have substrate in their active sites. Adding more substrate makes no difference to the rate when this point is reached.

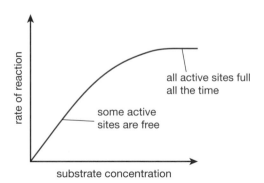

all active sites full all the time

some active sites are free

This sketch graph shows the rate of a reaction at different substrate concentrations. All other conditions were kept constant.

Questions

1 What are enzymes made of?

2 The enzyme amylase breaks down starch to sugar (maltose). The enzyme catalase breaks down hydrogen peroxide to water and oxygen. Explain why catalase cannot break down starch.

3 Explain why enzyme-catalysed reactions stop at very high temperatures.

4 What is meant by an enzyme's optimum temperature?

5 Enzymes in food and enzymes from bacteria and fungi make food decay. Why does food stay fresher for longer in a refrigerator than it does at room temperature?

Key words

➤ rate
➤ optimum
➤ denatured

C: The rate of photosynthesis

Find out about

- how temperature, light intensity, and carbon dioxide concentration affect the rate of photosynthesis
- **H** limiting factors

Key words

➤ light intensity
H ➤ inverse square law

The conditions inside the greenhouse in the photograph are very carefully controlled to provide the optimum conditions for photosynthesis. This enables the tomato plants to make glucose at their fastest rate, so they are growing quickly. The farmer plans all this so that the yield from the tomato plants will be as high as possible. Yield is the amount of product the farmer has to sell.

The effect of temperature

The greenhouse is kept warm at 26 °C. This is the optimum temperature for photosynthesis to take place in these tomato plants. This is because 26 °C is the optimum temperature for the enzymes used in photosynthesis in the tomato plant cells.

Some plants don't grow outdoors in the UK because the temperature is too cold. They are better adapted for life in hot climates.

The effect of light intensity

Light is needed for photosynthesis, so increasing the amount of light a plant receives increases the rate of photosynthesis.

The diagram below shows an experiment to investigate how changing **light intensity** affects the rate of photosynthesis in pondweed.

The temperature falls as you climb a mountain (the temperature decreases as the height increases). Above a certain height, large plants are absent because it is too cold for them to photosynthesise effectively.

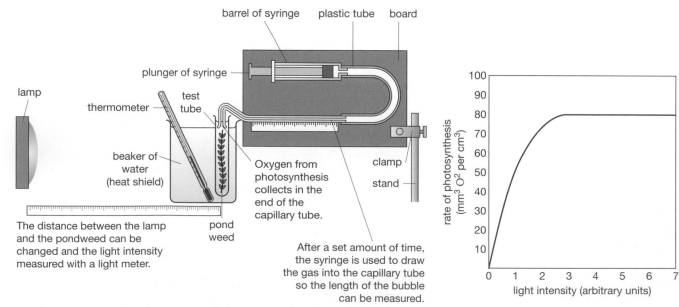

The distance between the lamp and the pondweed can be changed and the light intensity measured with a light meter.

Oxygen from photosynthesis collects in the end of the capillary tube.

After a set amount of time, the syringe is used to draw the gas into the capillary tube so the length of the bubble can be measured.

This experiment investigates how changing the light intensity affects the rate of photosynthesis. The light intensity is increased by moving the lamp closer to the beaker.

The results from the experiment are shown in the graph. The graph shows that:

● at low light intensities, increasing the amount of light increases the rate of photosynthesis

● at a certain point, increasing the amount of light stops having an effect on the rate of photosynthesis.

H The effect of distance on light intensity

Light waves spread out as they move away from a point light source such as a lamp. This means that further away from the lamp, the same amount of light is spread out over a larger area. So, each unit of area receives a smaller proportion of the total light – in other words, the light intensity decreases.

The relative light intensity (**LI**) at any distance (**d**) from a light source is inversely proportional to the square of the distance from the light source. This is called the **inverse square law**. It is an example of a mathematical model. This relationship can be written:

$$LI \propto \frac{1}{d^2}$$

Because the two variables are *inversely* proportional to one another, light intensity decreases as the distance increases.

The relationship between light intensity and distance is illustrated in the diagram in the margin. In the diagram, each small square is one unit of area. When the distance (**d**) from the light source is doubled, the light is spread over four units of area. So the light intensity in each unit of area is $\frac{1}{4}$ (or $\frac{1}{2^2}$). When the distance is tripled, the light is spread over nine units of area, so the light intensity in each unit of area is $\frac{1}{9}$ (or $\frac{1}{3^2}$).

This relationship helps to explain why the rate of photosynthesis (which is affected by light intensity) is lower when a plant is further away from its light source.

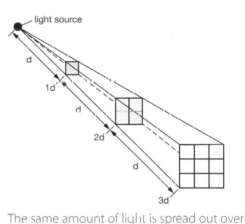

The same amount of light is spread out over a larger area as the distance (**d**) from a light source increases.

Worked example: The effect of distance on light intensity

Tamal places a plant 1 m from a lamp. The lamp is the only source of light. He works out the rate of photosynthesis in the plant. He then changes the distance to 2 m from the lamp, and works out the rate of photosynthesis. The temperature and the amount of carbon dioxide remain constant.

Use the inverse square law to explain why the rate of photosynthesis is lower at 2 m from the lamp than at 1 m.

Step 1: Explain which factor will affect the rate of photosynthesis in this experiment.

The rate of photosynthesis depends on the light intensity.

Step 2: Write down the relationship between relative light intensity and distance from the light source (the inverse square law).

The light intensity is inversely proportional to the square of the distance from the light source.

Step 3: Explain how the light intensity changes in Tamal's experiment.

When the distance is doubled from 1 m to 2 m, the light intensity is reduced to $\frac{1}{4}$ (which is $\frac{1}{2^2}$).

Answer:

The rate of photosynthesis is lower at 2 m from the light source because the light intensity at 2 m is only $\frac{1}{4}$ of the light intensity at 1 m.

The rate of photosynthesis cannot increase forever

The graph on the previous page shows that increasing the light intensity above a certain amount does not increase the rate of photosynthesis any further.

Photosynthesis needs more than just light. Extra light makes no difference to the rate of photosynthesis if the plant does not have enough carbon dioxide, water, or chlorophyll. The temperature must also be high enough for photosynthesis reactions to take place. Increasing the light intensity stops having an effect on the rate of photosynthesis when one of these other factors is in short supply. This factor is called the **limiting factor**.

Limiting factors

The graph below shows the effect of increasing light intensity on the rate of photosynthesis in tomato plants at two different carbon dioxide (CO_2) concentrations. At 0.04% CO_2, more light increases the rate of photosynthesis up to the point labelled A. After point A, extra light does not increase the rate so a factor other than light must be limiting it. Increasing the CO_2 level to 0.1% (as shown by the blue line on the graph) enables the rate of photosynthesis to increase, so CO_2 must have been the limiting factor. But the rate is again limited at the point labeled B, as either CO_2 or another factor becomes limiting.

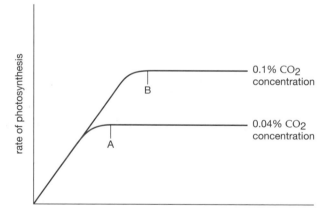

At the higher carbon dioxide concentration, photosynthesis in tomato plants happens at a faster rate but the rate still levels off. A factor is still limiting the rate of photosynthesis.

Questions

H

1   Write down four factors that can affect the rate of photosynthesis.

2   Explain what is meant by a limiting factor.

3   Suggest a factor that could be limiting the growth of bluebells on a woodland floor in spring.

4   Think about the tomato greenhouse from the start of this section. The concentration of carbon dioxide in the air inside the greenhouse is 0.04%. The farmer is planning to increase this concentration. He hopes this will increase the yield of tomatoes.

   What should the farmer consider when deciding on the concentration of carbon dioxide to use inside the greenhouse?

A: Moving molecules

Photosynthesis takes place in cells in plants. The process uses carbon dioxide and water, and produces oxygen. Plants also need nitrogen – they use it to make proteins for growth. How do molecules of these substances get into and out of plant cells?

Molecules move into and out of cells all the time. All living cells need a constant supply of substances for chemical reactions and they need to remove waste substances. Molecules of these substances move into and out of cells using three important processes. In this section we will discuss these processes in plant cells, but it's important to remember that they happen in all cells – animal, plant, and bacterial.

Diffusion

Molecules of gases and liquids move about randomly, all the time. They collide with one another and change direction. This makes them spread out. This is **diffusion**.

Molecules diffuse out of a tea bag when you make a cup of tea.

Water has just been poured onto this tea bag. Dissolved molecules from the tea are concentrated close to the tea bag. There are few tea molecules in the rest of the water.

30 seconds later: tea molecules are beginning to move from where they are concentrated into regions where they are less concentrated.

3 minutes later: the tea molecules are spread evenly throughout the water.

The **net movement** of molecules by diffusion is from an area of their high concentration to areas of their lower concentration. This means that, overall, more molecules move away from where they are more concentrated than move the other way.

Diffusion does not need any more energy than is in the molecules already. It is a **passive** process.

Key words

➤ diffusion
➤ net movement
➤ passive

Key words

➤ stomata
➤ gas exchange
➤ osmosis
➤ partially permeable membrane
➤ root hair cell

Diffusion and gas exchange in leaves

The surface of a leaf contains thousands of tiny holes called **stomata**. There are usually more stomata on the underside of a leaf than on the topside. Stomata allow carbon dioxide to enter the leaf for photosynthesis. Molecules of carbon dioxide diffuse from a region of higher concentration (the atmosphere) to one where the concentration is lower (spaces inside the leaf filled with air). Stomata also allow molecules of oxygen made by photosynthesis to diffuse out.

The movement of carbon dioxide and oxygen into and out of plants is an example of **gas exchange**.

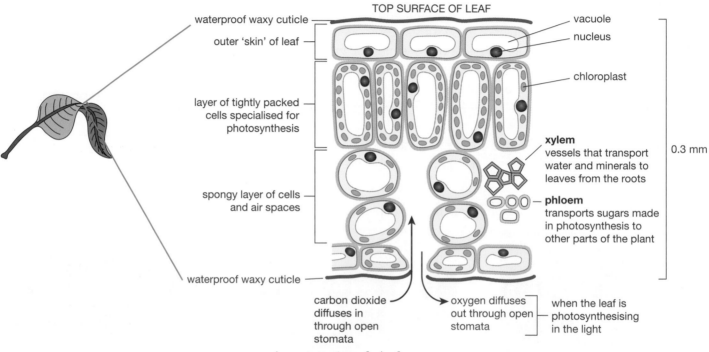

A cross-section of a leaf.

Location	Carbon dioxide	Oxygen	Result
cells in the leaf	Carbon dioxide is used in cells for photosynthesis. There is a lower concentration of carbon dioxide in cells in the leaf.	Oxygen is made by photosynthesis. A higher concentration of oxygen builds up in cells of the leaf.	Carbon dioxide diffuses through the stomata and into cells in the leaf to supply photosynthesis.
air surrounding the leaf	There is a higher concentration of carbon dioxide in the air.	There is a lower concentration of oxygen in the air.	Oxygen diffuses out of the cells and through the stomata.

A summary of gas exchange in a photosynthesising leaf.

Osmosis

When water molecules move by diffusion we call this **osmosis**. Osmosis is a specific type of diffusion. Water for photosynthesis and other processes enters cells by osmosis.

The cell membrane is important in osmosis. Cell membranes let some molecules through but block others. Tiny channels in the membrane allow small molecules, such as water and gases, to travel through them. Larger molecules are too big and cannot get through. So a cell membrane is a **partially permeable membrane**.

The two diagrams show how water molecules move through a partially permeable membrane by osmosis. In this example, the water molecules can get through the channels in the membrane, but the glucose molecules are too big to get through.

Key

 partially permeable membrane allows some molecules through and acts as a barrier to others

○ glucose molecule

○ water molecule

water molecules associated with glucose molecule (these molecules are not free to move by osmosis)

(*Note:* In these diagrams, the circles represent molecules, not individual atoms. Cell membranes are also made of molecules.)

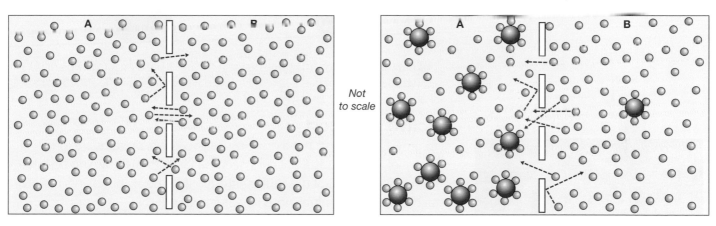

Not to scale

The effect of a partially permeable membrane. In the first example, the membrane is separating water molecules. All of the molecules move at random. The number of molecules moving from side A to side B is equal to the number moving from side B to side A. In the second example, the membrane is separating two glucose solutions. There are more free water molecules and fewer glucose molecules on side B. Water molecules that are free to move by osmosis are in higher concentration on side B. So there is a net movement of water from side B to side A.

Getting water from soil

Water enters a plant through the plant roots. Water molecules move across the membranes of root cells by osmosis. A **root hair cell** is adapted to increase the amount of water absorbed. Each of these cells has a structure that increases the surface area across which osmosis can occur. These structures are not actually hairs, but are extensions of the cell cytoplasm surrounded by the cell membrane and cell wall.

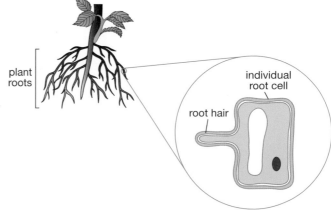

Root hair cells increase the surface area of a root. This enables the root to absorb more water from the soil.

The importance of osmosis

Plants do not have skeletons to give them support. Having cells that are just the right size and shape helps them to keep their structure. Osmosis is important in this. It drives the uptake of water by plant roots and determines how water passes from one cell to another throughout the whole plant.

If plant cells take in water, they bulge and become stretched. They are said to be **turgid**. Their strong cell wall prevents them from bursting. If plant cells lose too much water, they shrink. Plant cells should be fully hydrated to be healthy.

The photographs show the same plant. In the right-hand image, the plant is shown after it has not been watered for 10 days. Notice how the structure has changed as the plant cells have dried.

Active transport

To grow larger, plants need proteins. As you know, proteins are long chains of amino acids. Plant cells need nitrogen to make amino acids from glucose. Most of the Earth's nitrogen is in the air, but plants mainly take in nitrogen as nitrate ions in water from the soil.

Plants also need magnesium to make chlorophyll, and phosphates to make DNA. As proteins are used to build cells and make enzymes, nitrates are needed in the largest quantities.

Root hair cells absorb nitrate ions from the soil. However, nitrate ions are at a higher concentration inside the root cells than in the surrounding soil. So diffusion naturally moves nitrate ions out of the roots and into the soil. To overcome this, cells use a process called **active transport** to move nitrates from the soil into the roots.

When a cell needs to take in molecules that are in higher concentration inside the cell than outside, active transport is used. Energy is required to transport molecules across the cell membrane.

Fertiliser adds nitrate and phosphate ions to the soil. This ensures plants can absorb enough of these substances to grow.

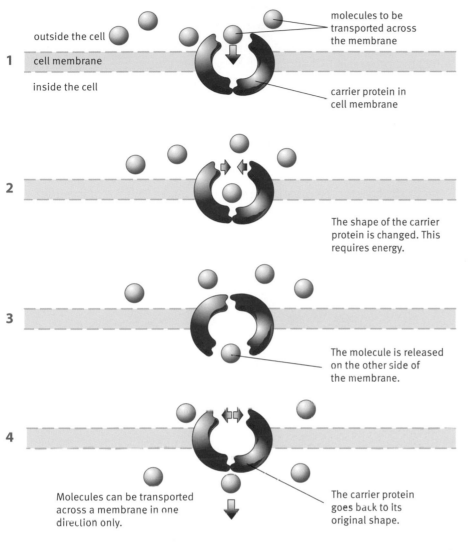

1 outside the cell

cell membrane

inside the cell

molecules to be transported across the membrane

carrier protein in cell membrane

2

The shape of the carrier protein is changed. This requires energy.

3

The molecule is released on the other side of the membrane.

4

Molecules can be transported across a membrane in one direction only.

The carrier protein goes back to its original shape.

Not to scale

Movement of molecules across a cell membrane by active transport.

Questions

1 Explain what is meant by a 'partially permeable membrane'.

2 Write down a definition of osmosis.

3 A student put a raisin (a dried grape) into a glass of water. She noticed that the raisin expanded and swelled. Explain this observation. Include a diagram to show the movement of molecules.

4 Glucose dissolves in water, but starch does not. Explain why glucose is stored as starch in plant cells.

5 Explain why cells sometimes need to use active transport.

6 Write down two main differences between diffusion and active transport.

B: Supply routes in plants

Animals have blood, blood vessels, and a heart to move substances around their body. Plants do not have this kind of circulatory system. Instead they have two separate transport systems. Substances can move both up and down the stem.

Plants have a tissue called **xylem** to transport water and minerals. These substances are moved from the roots up the stem to leaves and flowers.

Sugars, amino acids, and other substances are transported by a tissue called **phloem**. The movement of these substances is from a **source** (e.g., a leaf or storage organ) to a **sink** (e.g., a root or developing fruit). Substances can move both up and down the stem.

Xylem and phloem

Both xylem and phloem are found in the stem and the root.

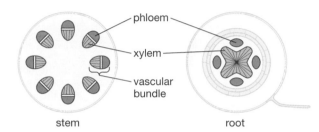

The arrangement of xylem and phloem in stem and root.

Xylem is made of dead cells. They are joined end to end to form a continuous tube with no obstructions. They have no cytoplasm. The walls are impermeable to water and contain lignin, which makes them tough.

Phloem is made of living cells. Like xylem, they are joined end to end to make a tube. But unlike xylem, they have cytoplasm.

Transpiration

Water is continually lost from the surface of a plant. It diffuses into the atmosphere through open stomata. This loss of water from the leaves is called **transpiration**.

Transpiration pulls water through the plant. Water is pulled through xylem tubes, up the stem from the roots. The water has minerals such as nitrates dissolved in it, which the plant needs for growth.

The rate of water uptake

The rate of water uptake by plants is affected by environmental factors. These include temperature, air movement, and light intensity:

- The rate increases as temperature increases. Water loss from the leaf cells, by evaporation, gets faster at higher temperatures. Water loss through the stomata by diffusion is also faster at higher temperatures.
- The rate increases as light intensity increases, because stomata open wider to get more carbon dioxide for photosynthesis.
- The rate increases with increased air movement. The moving air takes water vapour away from the leaf, making its surroundings drier. This increases the concentration gradient from leaf to atmosphere.

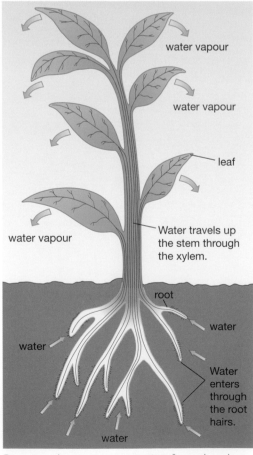

Diagram showing transpiration from the plant surface and the absorption of water through the roots.

Synoptic link

You can learn more about measuring the rate of water uptake by a plant using a potometer in B8K *Measuring the rate of water uptake by a plant.*

This flowering stem loses water by transpiration and blue-coloured water is absorbed from the vase to replace it. After a while, the blue dye can be seen in the white flower petals.

Translocation

Substances such as sugars are transported by **translocation**.

Translocation starts at the source (usually leaves) where sugars are made. Sugars are moved into the phloem by active transport. Water is then drawn into the phloem by osmosis. This causes an increase in pressure in the phloem tubes, which pushes the substances along the tubes. Finally, the sugar is unloaded at the sink (e.g., developing roots, stems, and flowers), where it is needed for cellular respiration and growth.

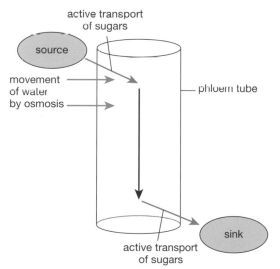

A summary of the movement of substances by translocation.

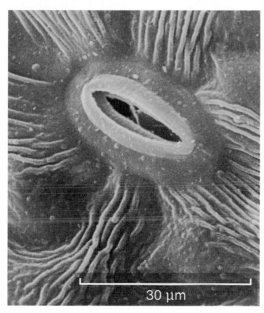

An image of a single stomata (properly referred to as a stoma when just talking about one of them), taken using an electron microscope. The hole is surrounded by a pair of guard cells. To close the hole, the guard cells take in water by osmosis and become swollen.

Questions

1 Make a table to show whether xylem or phloem transports each of the following:
   - water
   - minerals
   - sucrose
   - amino acids.

2 Explain the difference between translocation and transpiration.

3 Explain how osmosis is involved in translocation.

A: The fates of glucose

Find out about

- the importance of photosynthesis in producing food for life on Earth
- how glucose is used to make other carbohydrates, lipids, and proteins
- how organisms in an ecosystem are interdependent and how they compete

Producers make all their own food in the form of glucose. They do this by photosynthesis. Every food chain starts with a producer, so all living organisms depend on the ability of producers to make food. But before a producer is eaten by another organism, it uses the glucose it makes to grow and survive.

Inside the producer: using glucose

The glucose made by photosynthesis is used by plant cells in two ways.

1: Making other substances

A glucose molecule is made of carbon, hydrogen, and oxygen atoms. It is a type of sugar, which is a **carbohydrate**. Glucose can be used to make other carbohydrates. Two important carbohydrates in plants are cellulose and starch. They are both polymers of glucose. They are made up of thousands of glucose molecules bonded together.

Starch is a storage molecule. It can be converted back to glucose when needed. Starch is stored in starch grains in leaf cells. Some plants have special organs, such as the tubers of a potato, that have cells filled with starch.

Glucose is not just used to make other carbohydrates. It is also used to make **lipids** (including fats) and combined with nitrogen and sulfur to make **proteins**.

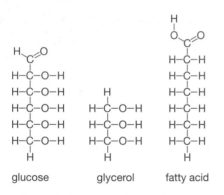

glucose   glycerol   fatty acid

The carbon (C), hydrogen (H), and oxygen (O) atoms in glucose can be used to make glycerol and fatty acids.

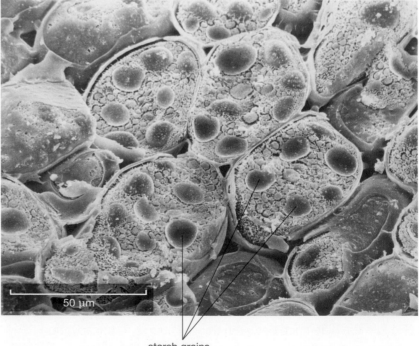

50 μm

starch grains

Some of the glucose from photosynthesis is made into starch. This is stored in starch grains inside plant cells.

The tubers of a potato plant store starch.

What happens to the glucose molecules?	What is made?	How is it used?
Many are joined together to make a polymer.	carbohydrates, such as starch and cellulose	Starch is used for storage. Cellulose is used for growth, including making cell walls.
The carbon, hydrogen, and oxygen atoms are used to make other molecules.	glycerol and fatty acids	Molecules of glycerol and fatty acids are combined to make lipids. Lipids are used for storage and making cell membranes.
The carbon, hydrogen, and oxygen atoms are combined with nitrogen and sulfur atoms to make other molecules.	amino acids	Many amino acids are joined together to make structural and functional proteins.

2: Cellular respiration

Glucose molecules are broken down by cellular respiration. This process transfers energy between stores, making it available for life processes such as chemical reactions in cells.

Passing it on: feeding relationships

A producer is an **autotroph** because it makes its own food (in the form of glucose). Organisms such as animals and fungi cannot make their own food. This kind of organism is a **consumer**, because it has to get its food from plants (or other consumers). A consumer is a **heterotroph**.

Grass is a producer. It is an autotroph that makes its own food by photosynthesis.

Producers use some of the glucose they make by photosynthesis to make **biomass**. Some of the biomass is transferred to other organisms when they eat the producers. An organism that eats producers is a primary consumer (e.g., a herbivore). Some of the biomass is transferred again when a secondary consumer (e.g., a carnivore) eats primary consumers.

Cellulose is the most abundant organic polymer on Earth. Wood is 50% cellulose, while cotton fibre is 90% cellulose.

Synoptic link

You can learn more about cellular respiration in B4.1 *What happens during cellular respiration?*

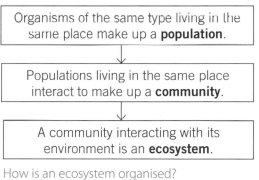

> Organisms of the same type living in the same place make up a **population**.

↓

> Populations living in the same place interact to make up a **community**.

↓

> A community interacting with its environment is an **ecosystem**.

How is an ecosystem organised?

Key words

➤ carbohydrate
➤ lipid
➤ protein
➤ autotroph
➤ consumer
➤ heterotroph
➤ population
➤ community
➤ ecosystem
➤ biomass

Models of feeding relationships: food chains and food webs

A **food chain** shows one route that biomass takes through an ecosystem. In this case, biomass is transferred from a producer through four levels of consumers. The plants make food for all the other organisms in the chain.

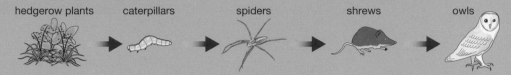

Only one organism is shown at each level, but each organism represents a population. We cannot tell from the food chain how big each population is. We don't know how many caterpillars are eaten by each spider, and so on. We don't know whether each population also appears in other food chains.

Many organisms eat, and are eaten by, more than one thing. A better model of feeding relationships in an ecosystem is a **food web**. It shows multiple food chains and how they interact.

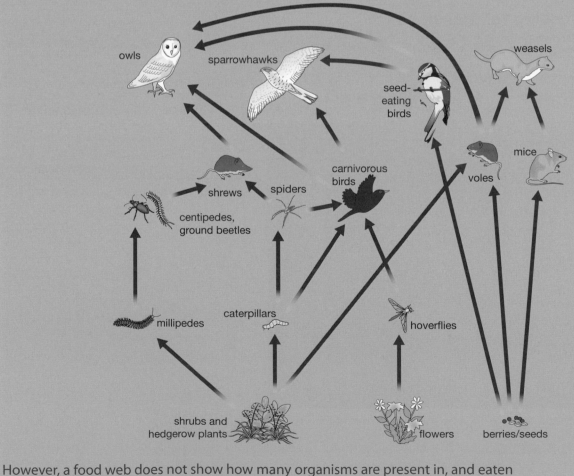

However, a food web does not show how many organisms are present in, and eaten from, each population. Also, it does not tell us how much biomass is transferred.

A food web tells us about the **interdependence** of populations of organisms. For example, we can use it to predict how a change in the size of one population would affect other populations in the same community.

A food web also shows how populations of organisms are in **competition** for food. For example, seed-eating birds, mice, and voles all compete for berries and seeds.

A food web only gives information about competition for food, but organisms compete for many resources, including:

● food

● water

● space

● light

● shelter

● mates

● pollinators

● seed dispersers.

When a resource is in short supply, the size of a population is limited by how well the organisms compete for the resource.

> **Key words**
> ➤ food chain
> ➤ food web
> ➤ interdependence
> ➤ competition

Humans are consumers. We eat and digest biomass from other organisms.

Inside the consumer

Consumers digest biomass from the organisms they eat. Digestion breaks down the carbohydrates, lipids, and proteins in the biomass. Carbohydrates are broken down into sugars (such as glucose). Lipids are broken down into fatty acids and glycerol. Proteins are broken down into amino acids. These products of digestion are absorbed into the blood. They are then transported to cells, and used to build biomass in the consumer. Consumers get their supply of carbon and nitrogen by eating other organisms and digesting the biomass.

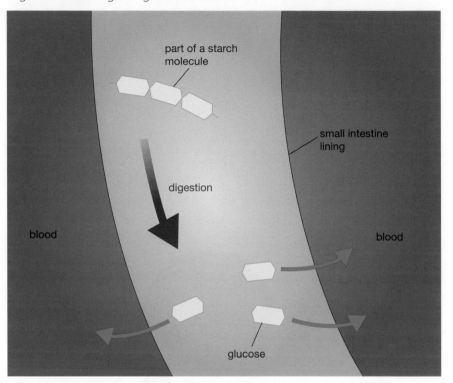

The biomass eaten by a consumer may contain starch. Starch is broken down into glucose molecules by digestion in the small intestine. The glucose molecules are small enough to be absorbed into the blood.

Questions

1 Name two things that:
   a plant species may compete for
   b animal species may compete for.

2 Look at the food web in the model box.
   a Name two different animals competing for the same food source.
   b A disease kills all the flowering plants. Explain what happens to the number of hoverflies.
   c Mink move into the habitat. They eat voles.
      i The number of mice decreases. Explain why.
      ii Explain what would happen to the number of caterpillars.

3 Explain why protein synthesis depends on the processes of photosynthesis and active transport.

B: Cycling substances

The substances that make your body are only on loan. They enter your body in the food you eat and the liquids you drink. There may be some substances in you right now, parts of which you will breathe out into the atmosphere tomorrow. Some of the water that you contain will end up in the toilet. When you die, all the substances in your body will leave it and re-enter the ecosystem when your body decomposes (decays).

The carbon cycle

When you breathe carbon dioxide into the atmosphere, it is entering an **abiotic** (non-living) part of the ecosystem. If, later, a plant absorbs it in photosynthesis, it is entering a **biotic** (living) part of the ecosystem. Substances cycle through the abiotic and biotic components of an ecosystem.

Carbon is important to living things because the substances from which they are made are all carbon compounds. All the organisms on Earth are carbon-based life forms.

Carbon cycles through the biotic and abiotic components of an ecosystem. The process of photosynthesis puts carbon into living things. In this process, carbon moves from the atmosphere (an abiotic store) into carbon compounds (initially glucose) in living things. This process is called carbon fixation. The carbon may stay in living things for just a few seconds or for many years. Eventually, other processes remove carbon from the living things and put it back into the abiotic stores (including the atmosphere, rocks, and fossil fuels).

The stages in the **carbon cycle** are shown in the diagram.

Find out about

- the abiotic and biotic components of an ecosystem
- the cycling of carbon and water through an ecosystem
- the role of decomposers

Key words

➤ abiotic
➤ biotic
➤ carbon cycle

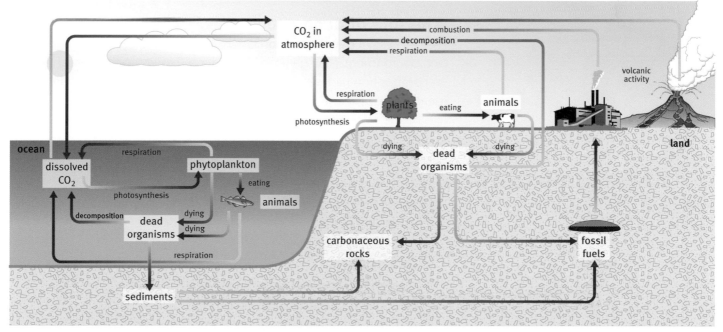

In the carbon cycle, one store of carbon is in the atmosphere as carbon dioxide. The fixation process is photosynthesis. The carbon, first fixed into glucose in plants, is made into starch, cellulose, proteins, and many other substances. It is then passed on to the herbivores that eat the plants. Carnivores get the carbon by eating herbivores. All these living things produce carbon dioxide in a process called respiration. During respiration they gain energy from the food and release carbon dioxide back into the atmospheric store.

Key words

➤ water cycle
➤ decomposition
➤ greenhouse gas

The water cycle

Cells cannot function without water; it is the solvent in cell cytoplasm, blood plasma, urine, sweat, saliva, and tears. It also keeps cell contents at the correct concentration for chemical reactions and other life processes to occur. Like carbon, water cycles through living and non-living things.

The diagram shows stages in the **water cycle**.

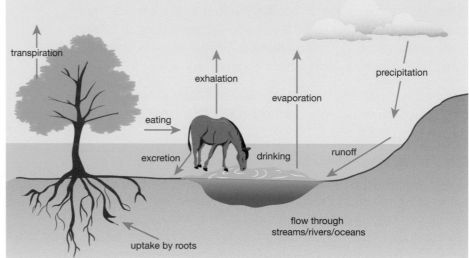

This dead bird is being decomposed.

Plants take in water by their roots. Animals take in water by drinking it and with their food. They both lose water in respiration. Plants lose most from their leaves; animals lose it via various organs, such as the kidney in mammals. Much water simply cycles as water through the abiotic and biotic components of the ecosystem.

The role of decomposers

The process of **decomposition** is very important. It returns minerals to the environment. New generations of living things can then use them. If decomposition had never happened, the world would be miles deep in dead bodies. No one would know, however, because there would be no life. All the essential substances would be locked up in dead bodies.

Microorganisms carry out decomposition using enzymes. All microorganisms need water to survive. Many also need oxygen. In places where there is no oxygen some decomposition can happen. It is called anaerobic decomposition and produces methane. Methane is a **greenhouse gas** with a much greater greenhouse effect than carbon dioxide.

Questions

1 Describe two ways in which carbon is:
   a added to the atmosphere
   b taken out of the atmosphere.

2 In which processes in the water cycle:
   a does water move by osmosis?
   b does water change state from a gas (water vapour) to a liquid?

A: Organisms and their habitats

Do plants just grow where we plant them? That may be the case in gardens and parks, but not in natural ecosystems. Some plants grow in shade while others need bright light. Some plants need plenty of water and others can survive in deserts. All plants need minerals, water, and light in just the right amounts to be able to grow well.

Plants and animals usually live in places where the conditions enable them to thrive. Species are adapted to survive in the **habitat** in which they live. Many different organisms have evolved to take advantage of the great variety of habitats all over the Earth.

Different conditions within an ecosystem

Conditions are not the same in all parts of an ecosystem. Conditions such as light intensity, temperature, the availability of water, and soil pH will be different in various parts of an ecosystem. This creates different habitats, capable of supporting different organisms. This affects the **distribution** of organisms (where they are found). It also affects their **abundance** (how many there are in an area).

Pollution of an ecosystem

Bioaccumulation

Certain substances may enter an ecosystem and be poisonous (toxic) to living things. In some cases the substance is not harmful in small amounts, but it can become concentrated by living things themselves along a food chain. This is called **bioaccumulation**.

Substances that are not broken down by living organisms can bioaccumulate in a food chain. The grass absorbs a substance, for example an insecticide, from the soil. There is not enough of the substance in each blade of grass for it to reach toxic concentration. But each mouse eats many blades of grass, and each owl eats many mice. At each trophic level, the amount of insecticide ingested remains the same, but there are fewer organisms so the amount of insecticide per organism is greater. It may reach fatal concentration in the owl.

Find out about

- how abiotic and biotic factors affect the distribution and abundance of organisms

Key words

- ➤ habitat
- ➤ distribution
- ➤ abundance
- ➤ bioaccumulation

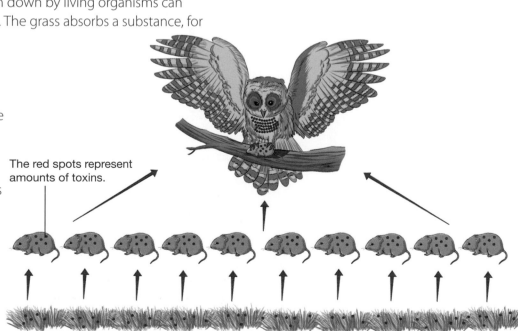

The red spots represent amounts of toxins.

Eutrophication

Sometimes, problems can be caused by too much of a certain substance, even though it is not toxic. When farmers add fertilisers containing nitrate and phosphate to their fields this can cause problems. The fertilisers are washed off by rain into streams, rivers, and lakes. In the water, the fertiliser causes algae to grow very quickly. The algae block light from the water plants below. The water plants cannot photosynthesise, so they die. The algae soon die as well. Bacteria decompose the dead plants and algae, using up lots of oxygen. The oxygen levels in the water go down quickly, causing fish and other animals to die. They are also then decomposed – increasing the problem. The process is called **eutrophication**. The same problem can be caused by putting sewage into water, as sewage contains nitrogen compounds.

Algae growing on the river Thames.

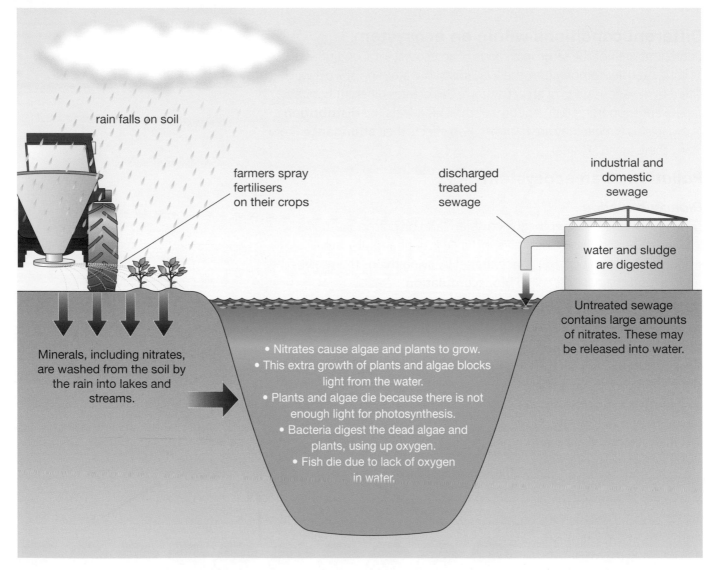

rain falls on soil

farmers spray fertilisers on their crops

discharged treated sewage

industrial and domestic sewage

water and sludge are digested

Minerals, including nitrates, are washed from the soil by the rain into lakes and streams.

- Nitrates cause algae and plants to grow.
- This extra growth of plants and algae blocks light from the water.
- Plants and algae die because there is not enough light for photosynthesis.
- Bacteria digest the dead algae and plants, using up oxygen.
- Fish die due to lack of oxygen in water.

Untreated sewage contains large amounts of nitrates. These may be released into water.

Too much fertiliser or sewage can lead to eutrophication and the death of fish and insects in freshwater or the sea.

Changing conditions

If the temperature changes or a toxic substance enters an ecosystem, this might lead to a change in the populations in a community. Because of the way all the living things in a community are interdependent, if one population changes, others will too.

Look at the food web. The population of berries and seeds may go down owing to a cold spring stopping good pollination. This would lead to a fall in populations of voles, mice, and seed-eating birds. Because weasels and owls feed on these, their populations would fall too. However, owls might start eating more carnivorous birds and cause a decline in their populations. If you look carefully you will see that everything in the food web would probably be affected by a cold spring leading to a reduction in berries and seeds.

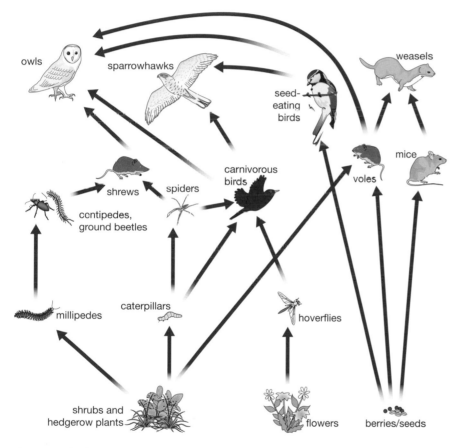

The arrival of new predators, pathogens, competitors, or toxic substances can all lead to similar changes across the food web.

Questions

1 Write down three things that a plant needs in its habitat.

2 Draw a flowchart to summarise the process of eutrophication.

3 A farmer sprays an insecticide on her crops. The insecticide kills insects. Voles and owls live in the same community. The insecticide is not harmful to owls at the concentration used. Explain how the insecticide could become toxic to the owls.

B: Investigating different habitats

Find out about

● ways to investigate the distribution and abundance of organisms in an ecosystem

● working out the number of organisms in an area

Key words

➤ representative sample
➤ quadrat
➤ identification key
➤ transect
➤ abiotic factors
➤ biotic factors

This student is using a quadrat to record the plants growing in a 0.5 m × 0.5 m square of playing field.

Investigating the distribution and abundance of organisms usually involves doing fieldwork. It is often not possible to count every organism in a population or to survey the whole of a large area such as a field. Instead, a sample is studied. Enough measurements must be taken to get a **representative sample** of the population or area.

Quadrats and transects

A **quadrat** is used to investigate the abundance of non-moving or slowly moving organisms in different parts of an ecosystem. For example, the abundance of different plant species could be measured using a quadrat. A quadrat is a square frame (usually with sides of 0.5 m or 1 m), which is divided into a grid to make counting individual organisms (or estimating percentage cover) easier. The size of quadrat that is chosen depends on the size of the plants and also the area that needs to be surveyed.

The quadrat is placed on the ground. Recording the species found inside the quadrat involves accurate identification of each species. An **identification key** can help. The number of individuals of each species is counted, or for grasses and bare soil the percentage cover is calculated.

To sample a large area or to compare samples from two different areas, quadrats should be placed at random. This helps to make sure the sample is a representative sample of the area.

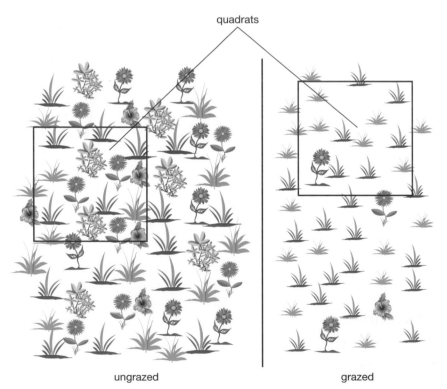

Quadrats placed randomly to investigate the distribution and abundance of plant species in grazed and ungrazed areas.

Sometimes samples are taken at regular intervals along a straight line called a **transect**. This provides a systematic sample of a small area. It is used to investigate how the types of plants change gradually from one area to another, for example, when moving from the shaded part of a wood into an open field.

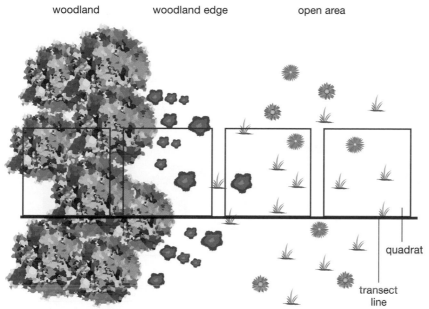

A transect with quadrats used to investigate distribution and abundance of plant species on a woodland edge.

Shady woodland	Middle of field
bare soil: 40% coverage	bare soil: 0% coverage
grass: 30% coverage	grass: 100% coverage
bluebell plants: six	bluebell plants: zero

Results from two different quadrats.

Explaining the distribution and abundance of organisms

To help explain why organisms grow in particular locations, various factors can be measured. These can include **abiotic factors** such as temperature, light intensity, soil pH, the availability of water, and levels of pollution. These factors can be measured in each quadrat, or at regular intervals along a transect line, using appropriate apparatus.

Biotic factors (such as competitors, predators, and pathogens) should also be taken into account.

Synoptic link

You can learn more about the techniques used to investigate habitats in B8I *Investigating the distribution and abundance of organisms* and in B8J *Measuring abiotic factors in an ecosystem.*

Estimating animal population size

The size of a population of animals can be estimated by taking samples.

A useful sampling method is **capture-mark-recapture**. In this method:

- A sample of individuals of one species is collected from an area.
- The individuals are marked so that they can be recognised.
- The marked individuals are released.
- Later, a second sample of individuals is collected.
- The number of marked individuals in the second sample is counted.
- An equation is used to estimate the population size.

This bird has been marked using a small metal ring around its leg.

This method assumes that the population size does not change between the first and second sample. For example, it assumes that no individuals move into or out of the area, and that none are born or die. It also assumes that the marks are not lost from any individuals.

This snail has been marked using a number painted onto its shell.

Worked example: Capture-mark-recapture

A student counts 10 snails in a quadrat. He marks each snail's shell using a dot of paint. One week later he returns, and counts 14 snails in the quadrat. Seven of these snails have marks.

Estimate the size of the snail population.

The following equation is used to estimate the population size from capture-mark-recapture samples:

$$\text{estimated population size} = \frac{\text{number of individuals given marks} \times \text{number of individuals recaptured}}{\text{number of recaptured individuals that have a mark}}$$

Step 1: Substitute the data into the equation.

$$\text{estimated population size} = \frac{10 \times 14}{7}$$

Step 2: Calculate the estimated population size.

$$\text{estimated population size} = \frac{140}{7}$$

Answer:

estimated population size = 20

Indicator species

Some species are only found when there is no pollution. Others are only found when there is pollution. These kinds of species can be used as **indicator species**. They help us to make conclusions about the amount of pollution present in parts of an ecosystem.

Kick sampling in a river.

Indicator species in rivers help us to measure how clean the water is (the water quality). One method of collecting indicator species is **kick sampling**:

- A net is held in the flowing river.
- The river bed upstream of the net is kicked for a fixed length of time (at least 30 seconds).
- The contents of the net are tipped into a white plastic tray.
- Organisms are identified and counted.

The number and types of organisms found give an indication of the water quality. For example, mayfly nymphs are only found in cleaner water. The more mayfly nymphs found, the cleaner the water. Mayfly nymphs are one example of an indicator species.

Questions

1 List, under two headings, which of the following are abiotic and which are biotic.

    predators   light intensity   pH   temperature   wind speed   competitors

2 Suggest why the capture-mark-recapture method may not give an accurate estimate of the size of an animal population.

3 The following is a list of investigations done by students on a field trip.

Suggest which sampling techniques you would use for each investigation.

- Investigating the change in living organisms from land down to the edge of the sea when the tide is out.
- Investigating the plant species present in mowed and unmowed meadows.
- Investigating the water quality in two different places in a stream.
- Comparing the size of the snail population in a rubbish dump and a nearby woodland.

Science explanations

B3 Living together – food and ecosystems

In this chapter you have learnt about ways in which organisms interact with one another and their environment, how life depends upon these interactions, and how survival can be endangered by natural and man-made environmental changes.

You should know:

- about the two main stages of photosynthesis, including the inputs and outputs of each stage, and the need for light in the first stage
- that photosynthesis is an endothermic process
- about the role of chloroplasts in photosynthesis
- how enzymes work, using the lock-and-key model
- why substrate concentration, temperature, and pH affect the rate of enzyme-catalysed reactions
- why carbon dioxide concentration, temperature, and light intensity affect the rate of photosynthesis, **H** and how these factors interact to become limiting
- how substances are transported into and out of cells by diffusion, osmosis, and active transport
- about partially permeable cell membranes
- how water and mineral ions are taken up by plants, including the role of root hair cells
- how xylem and phloem are adapted to transport substances in plants
- about the roles of transpiration and translocation in plants
- about the structure and function of stomata
- how light intensity, air movement, and temperature affect the rate of water uptake by plants
- what carbohydrates, lipids, and proteins are made of (and broken down into)
- that producers are the source of biomass for food chains, and how biomass is transferred between organisms in an ecosystem
- how feeding relationships can be modelled using food webs
- how substances such as carbon and water are cycled through ecosystems
- about the role of microorganisms in cycling substances
- how abiotic and biotic factors affect communities, including environmental conditions, toxic substances, availability of food, and other resources, and the presence of predators and pathogens
- how to investigate the distribution and abundance of organisms in an ecosystem.

During photosynthesis, energy from the Sun is used to top up chemical stores of energy in leaves.

Ideas about Science

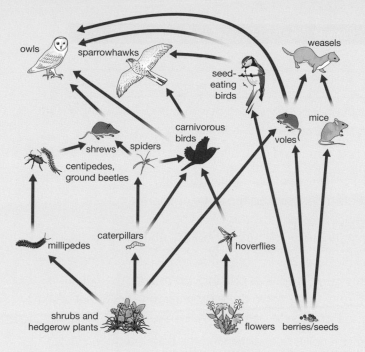

A food web is a model of the feeding relationships between populations in a community.

Models are used in science to help explain complex interactions. The lock-and-key model explains the interaction between an enzyme and its specific substrate. Food chains and food webs help explain feeding relationships within a community of interacting populations. These models can also help us make predictions about how changes in factors will affect outcomes. For example, we can predict that higher temperatures will change the shape of an enzyme's active site, which will decrease the rate of reaction. Or we can predict how a change in the size of a population will affect others in the same food web.

For each of these different models you should be able to:

- describe which bits of information or data are included in the model
- use the model to explain how an enzyme and substrate interact, or how populations in a community interact
- use the model to make a prediction about how changes in factors will affect how the parts interact, and the outcome of this
- identify limitations of the model.

Science and technology provide people with many things that they value, and that enhance their quality of life. However, some applications of science can have unintended and undesirable impacts on ecosystems. For example, fertilisers used in farming to increase the yield of food can be washed into lakes and rivers, causing eutrophication. Insecticides used to protect food crops can bioaccumulate in food chains and reach toxic levels in consumers.

You should be able to:

- identify unintended impacts of human activity on the environment
- describe ways in which we can reduce these effects.

Eutrophication is an unintended impact of human activity on the environment.

B3 Review questions

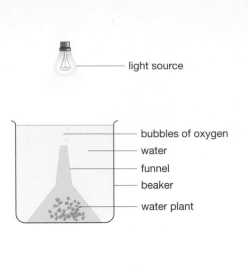

light source

bubbles of oxygen
water
funnel
beaker
water plant

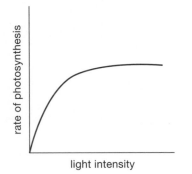

rate of photosynthesis

light intensity

1   Photosynthesis takes place in green plants.

   a   Copy and complete the table to summarise the process
       of photosynthesis.

Stage	Inputs	Outputs
1		
2		

   b   The rate of photosynthesis can be measured. A student sets up
       their apparatus, as shown in the diagram in the margin.

       Explain how you could estimate the rate of photosynthesis using
       this apparatus.

   c   A different student sets up the same apparatus but also includes
       a gas syringe. Explain how this could be used to give a more
       accurate estimate of the rate of photosynthesis.

   H d   A farmer grows plants in a greenhouse. He wants to make sure
       the plants grow as large as possible. He uses lamps to increase the
       light intensity. Write down three other factors the farmer could
       change to increase the rate of photosynthesis.

   e   The farmer records how increasing the light intensity affects the
       rate of photosynthesis in the plants. The graph in the margin
       shows his results.

       Explain why the rate does not continue to increase as the light
       intensity increases.

2   Alex draws diagrams to show stages of a reaction involving an enzyme.

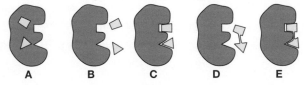

A   B   C   D   E

   The stages are not drawn in the correct order. One of the stages is
   not correct.

   Choose the correct stages and write them in the correct order.

   The final stage has been done for you.

			D

3   Starch synthase is an enzyme found in potato tubers, which grow in the
   UK. The enzyme joins glucose molecules together to store them as starch.
   Some students compared its rate of reaction at 5 °C, 20 °C, and 45 °C.

   Think about the environment that the potato grows in and predict
   what would happen if the students:

   a   raised the temperature of the tuber at 5 °C to 20 °C?

   b   raised the temperature of the tuber at 20 °C to 45 °C?

4 Molecules move across cell membranes by diffusion, osmosis, and active transport.

Describe and explain the differences between these three processes.

5 Xylem tissue is found in plants.

a Explain how the structure of xylem is adapted to its function.

b Explain how substances are moved through xylem.

6 Look at the food web.

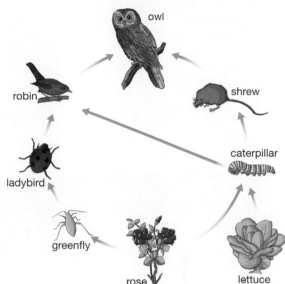

a Write down one producer from the food web.

b Write down one heterotroph from the food web.

c Write down one food chain from the food web.

d A pathogen kills all the rose plants. Explain what effects this could have on the other populations in the food web.

7 A food web is a type of model.

a Which statement correctly describes what the model represents?

A One route that biomass takes through an ecosystem.

B Feeding relationships within a population.

C A pyramid of numbers.

D Feeding relationships within a community.

b Like all models, a food web has limitations.

Which of the following bits of information are not included in a food web?

A Which organisms are consumers.

B How many organisms are present in each population.

C How many organisms are eaten from each population.

D How much biomass is transferred.

E Which organisms are autotrophs.

8 Describe the role of bacteria in the carbon cycle.

9 Inner Farne is an island off the northeast coast of England. A population of puffins lives on the island.

Describe how you would estimate the size of the puffin population.

B4 Using food and controlling growth

Why study cellular respiration and growth?

Look inside the cells of the largest animal or the smallest bacterium and you will find the same things happening. All living organisms need to get energy from glucose in the process of cellular respiration. They also need to repair damage and grow new cells. All organisms change as they grow and develop. Understanding these processes could help us to treat disease, and enables us to investigate the workings of all life on Earth.

What you already know

- All organisms are made of cells.
- All cells rely on cellular respiration to transfer energy for life processes.
- Aerobic respiration uses oxygen while anaerobic respiration does not.
- The inputs and outputs of aerobic and anaerobic respiration in animal cells, plant cells, and bacteria.
- How to use a light microscope to observe cells.
- Living things produce offspring of the same kind, but normally offspring vary and are not identical to their parents.
- Gametes and fertilisation are required for sexual reproduction in animals.
- The genome controls the function of cells.
- Risk factors for cancer include contact with substances called carcinogens, ionising radiation, particular genetic variants, and an unhealthy diet.
- Some applications of science can help improve our quality of life, but the benefits have to be weighed against the risks and ethical issues.

The Science

All life depends on the process of cellular respiration. It happens in all living cells, all of the time, and without it they would die. It's more than a simple chemical reaction, and different types of respiration are used in different conditions.

Of course we do more than just stay alive – we also grow and develop. This depends on cells dividing and becoming organised into tissues that do specific jobs in the body. Understanding and controlling development could help us develop new treatments for damage and disease.

Ideas about Science

How do we know what we know about cells? Some explanations were only developed because new technology helped us make new observations. Microscopes enable us to look inside cells and explain how the structures there help us to stay alive.

Other applications of science, such as stem cell technology, are often in the news. They offer great potential for medical treatments, but different people argue for and against their use. How can we weigh up the benefits, risks, and ethical issues to make a sensible decision?

A: Cellular respiration in animals

Find out about

- aerobic and anaerobic respiration in animal cells
- how mitochondria are related to cellular respiration
- why respiration can be described as an exothermic reaction

Your body is very demanding. Billions of cells carry out thousands of chemical reactions every second to keep you alive. Even when you sit still, tissues such as heart muscle never stop moving. You grow bigger and you constantly repair damage to your body. All of these processes, and many more, depend on **cellular respiration**.

Using biomass

Humans and other animals are consumers. When they eat other organisms, they digest most of the biomass from those organisms. Small molecules, including glucose, are absorbed from the gut and transported in the blood. Cells use the small molecules to build biomass in the animal. They also use some of the glucose for cellular respiration.

What is cellular respiration?

As you know, cellular respiration is an essential process. It happens in every cell in every living organism. A cell that is not carrying out cellular respiration is a dead cell.

Cellular respiration is a series of chemical reactions. These reactions make molecules of a substance called **ATP**. Your cells need a constant supply of ATP. It is needed for processes essential for life, including:

- muscle contraction
- active transport
- many chemical reactions (including the breakdown and making of carbohydrates, proteins, and lipids).

Even when you are asleep, every cell in your body is very busy. Cells and tissues carry out chemical reactions, they move molecules around, and they grow and repair themselves.

This woman is using ATP for many different things, including processing information in her brain, moving, maintaining a constant internal environment, and growing and repairing her tissues.

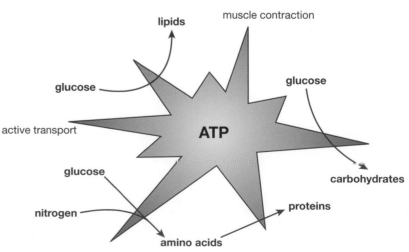

ATP made by cellular respiration is essential for many different life processes.

Many of the reactions in cellular respiration take place inside the **mitochondria** of cells. Mitochondria are organelles found in the cell cytoplasm.

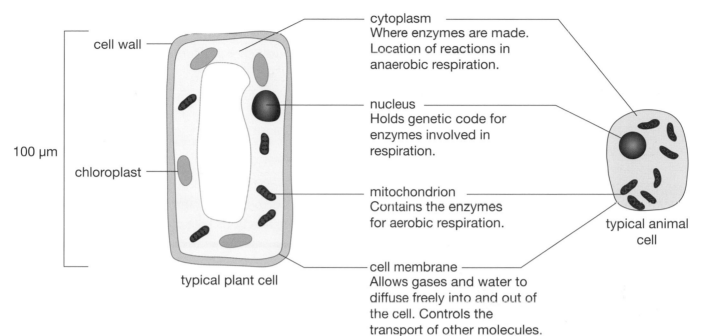

cell wall

100 μm

chloroplast

typical plant cell

cytoplasm
Where enzymes are made. Location of reactions in anaerobic respiration.

nucleus
Holds genetic code for enzymes involved in respiration.

mitochondrion
Contains the enzymes for aerobic respiration.

cell membrane
Allows gases and water to diffuse freely into and out of the cell. Controls the transport of other molecules.

typical animal cell

Mitochondria are found in plant and animal cells.

What is ATP?

Energy stores are involved in all processes in living organisms. ATP is a chemical store of energy. It is vital in living cells. In each molecule of ATP there are three phosphate groups. The 'T' in ATP stands for 'tri' (meaning three) and the 'P' stands for 'phosphate'.

Molecules of ATP are made during cellular respiration from molecules of a substance called ADP and phosphate ions. Each molecule of ADP contains two phosphate groups (the 'D' stands for 'di', meaning two).

ADP   phosphate   ATP

ATP is made during cellular respiration. It contains three phosphate groups. It is made from ADP, which contains two phosphate groups.

Many life processes convert ATP back into ADP. This means ATP is continuously recycled in living organisms. The human body contains about 250 g of ATP at any one time, but uses its own body weight in ATP every day.

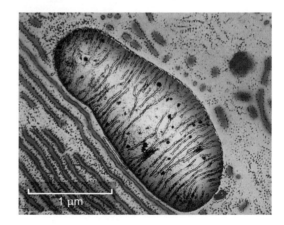

1 μm

This electron micrograph shows a single mitochondrion inside a cell. The mitochondrion has been coloured blue and yellow by a computer, to make it easier to distinguish it from the other cell structures.

Synoptic link

You can learn more about energy stores in P2.1A *What's the big energy picture?*

Key words

➤ cellular respiration
➤ ATP
➤ mitochondria
➤ aerobic respiration
➤ anaerobic respiration
➤ lactic acid
➤ exothermic

Models of aerobic respiration: equations

Aerobic respiration can be represented simply using a model such as a word equation or word summary. The model uses words to represent the reactants and products.

$$\text{glucose} + \text{oxygen} \longrightarrow \text{carbon dioxide} + \text{water}$$

Another model of aerobic respiration is a balanced chemical equation. It uses chemical formulae to represent the reactants and products.

$$C_6H_{12}O_6(aq) + 6O_2(g) \longrightarrow 6CO_2(g) + 6H_2O(l)$$

We can also show that ATP is made during the process:

$$\overset{\text{ADP} + \text{P} \quad \text{ATP}}{C_6H_{12}O_6(aq) + 6O_2(g) \longrightarrow 6CO_2(g) + 6H_2O(l)}$$

These equations are convenient ways of summing up the process of aerobic respiration. However, they miss out much of the detail. They suggest that only one chemical reaction is involved. In fact, aerobic respiration involves a number of chemical reactions. The equations do not tell us where the reactions take place.

The cells in a runner's muscles use two different types of cellular respiration.

Questions

1   a   Write down three processes in your body that need ATP.

   b   Explain why cellular respiration must happen continuously in all living cells.

2   Write bullet points to show the differences between aerobic and anaerobic respiration.

3   Steve is an athlete. He runs in an 800 m race. As he runs, his muscle cells use both aerobic and anaerobic respiration. Explain why Steve's muscle cells use both types of respiration when he runs the race.

Different types of cellular respiration in animals

Your cells can use two different types of cellular respiration:

● **aerobic respiration** – needs oxygen

● **anaerobic respiration** – does not need oxygen.

When an athlete exercises, their muscles need lots of ATP so they can contract. Most of this ATP comes from aerobic respiration. During aerobic respiration, glucose reacts with oxygen inside cells. At the end of the process, carbon dioxide and water are made as waste products.

When an athlete sprints, their muscles cannot get oxygen quickly enough for aerobic respiration. So they use anaerobic respiration, which does not need oxygen. This also happens when, for example, a predator runs after its prey, and the prey runs for its life.

In humans and other animals, anaerobic respiration can only be used for a short period of time. The process makes fewer molecules of ATP from each gram of glucose than aerobic respiration. It also makes a waste product called **lactic acid**. This is toxic in large amounts. If it builds up in muscles, it makes them feel tired and painful.

The word equation (word summary) for anaerobic respiration in animal cells is:

$$\text{glucose} \longrightarrow \text{lactic acid}$$

Cellular respiration and energy

Cellular respiration transfers energy between chemical stores. Energy is transferred from glucose and oxygen to ATP. Cellular respiration also warms its surroundings – it is an **exothermic** process. Some of the energy is transferred to the organism's tissues and its surroundings. The tissues and surroundings are thermal stores, and the energy in them increases.

Molecules of ATP are used by lots of different processes in cells. ATP is a chemical store of energy. Many life processes need energy from this chemical store.

B: Anaerobic respiration in plants and microorganisms

Anaerobic respiration in animal cells makes lactic acid. But in plants, yeast, and some bacteria it is different. In these organisms, anaerobic respiration makes ethanol and carbon dioxide. The word equation for this type of anaerobic respiration is:

$$\text{glucose} \longrightarrow \text{ethanol} + \text{carbon dioxide}$$

Anaerobic respiration in plant cells

When a seed starts to germinate, there is very little oxygen available to the cells inside the seed. By using anaerobic respiration, they can still release ATP to drive the chemical reactions that are used to start germination. The same happens when a plant lives in waterlogged soils; the roots find it hard to get oxygen and undertake anaerobic respiration instead.

Anaerobic respiration in microorganisms

Unlike animal and plant cells, bacteria do not have mitochondria. The reactions of cellular respiration in bacteria take place in the cell cytoplasm.

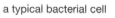

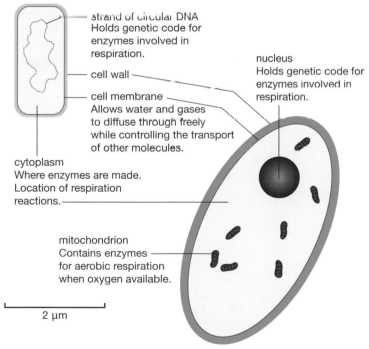

a typical bacterial cell

strand of circular DNA
Holds genetic code for enzymes involved in respiration.

cell wall

cell membrane
Allows water and gases to diffuse through freely while controlling the transport of other molecules.

nucleus
Holds genetic code for enzymes involved in respiration.

cytoplasm
Where enzymes are made. Location of respiration reactions.

mitochondrion
Contains enzymes for aerobic respiration when oxygen available.

2 μm

a typical yeast cell (fungus)

Structure of a typical bacterium and yeast cell.

Bacteria use anaerobic respiration in conditions of low (or no) oxygen, for example, inside an infected wound in the skin. However, anaerobic respiration in some bacteria and in yeast helps us to produce useful foods and fuels. For example, some bacteria are used to make cheese and yoghurt.

When yeast cells carry out anaerobic respiration, the process is often called **fermentation**.

Find out about

- anaerobic respiration in plant cells and microorganisms
- how the rate of cellular respiration can be calculated

Key word

➤ fermentation

Germinating seeds use anaerobic respiration.

This factory produces bioethanol to fuel cars.

Yeast uses sugars in the flour for anaerobic respiration. The carbon dioxide produced makes bread rise.

Anaerobic respiration in yeast is used in brewing to make wine and beer.

On a large scale, fermentation can produce bioethanol. This can be used to fuel vehicle engines. It is made from the sugars in plant material, such as sugar beet, maize, and wheat. Fuel from renewable sources is considered to be more sustainable and causes less global warming.

On a smaller scale, anaerobic respiration in yeast is useful for brewing and baking, producing bread, alcoholic drinks, and vinegar.

Calculating the rate of cellular respiration

The rate of cellular respiration can be calculated by following the formation of product. For example, we can measure how quickly carbon dioxide gas is made by anaerobic respiration (fermentation) in yeast. Look at the diagram. The conical flask contains 5 g of glucose, 50 cm^3 of warm water, and 1 g of yeast. The volume of carbon dioxide produced is recorded using the gas syringe.

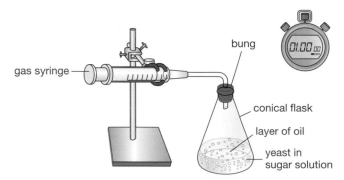

The results of the experiment are shown on the graph.

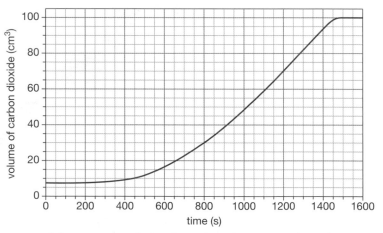

At the start of the reaction, the carbon dioxide was produced quite slowly. The volume of carbon dioxide only increased a small amount in the first 400 s. After 400 s the rate of production got much quicker, before slowing down again after about 1475 s.

We can calculate average rate of change in the volume of carbon dioxide between any two time points on the graph using the following equation:

$$\text{rate of change of volume (cm}^3\text{/s)} = \frac{\text{change in volume (cm}^3)}{\text{time taken for change (s)}}$$

We can use the average rate of change in the volume of carbon dioxide as our best estimate of the average rate of reaction of cellular respiration.

Worked example: Calculating the rate of cellular respiration

Use the graph to calculate the rate of cellular respiration in the yeast between 800 and 1200 seconds.

Step 1: Use crosses (x) to mark the two time points on the curve.

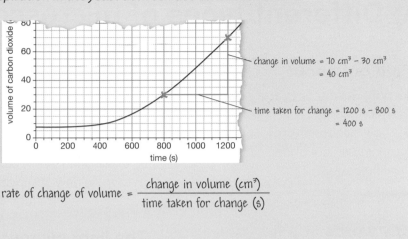

change in volume = 70 cm³ – 30 cm³
= 40 cm³

time taken for change = 1200 s – 800 s
= 400 s

Step 2: Mark on the graph the changes in values.

Step 3: Write down the equation you will use to calculate the average rate of change of volume.

$$\text{rate of change of volume} = \frac{\text{change in volume (cm}^3\text{)}}{\text{time taken for change (s)}}$$

Step 4: Substitute the quantities into the equation and calculate the average rate.

$$\text{rate of change of volume} = \frac{40 \text{ cm}^3}{400 \text{ s}}$$
$$= 0.1 \text{ cm}^3/\text{s}$$

Answer: The average rate of formation of carbon dioxide is 0.1 cm³/s between 800 s and 1200 s. This is the best estimate of the average rate of reaction of cellular respiration between these two time points.

Questions

1 Make a table to compare anaerobic respiration in animal cells and in yeast cells. You may want to think about the conditions under which they happen, the reactants, and the products.

2 Anaerobic respiration in animal cells makes a toxic product (lactic acid). Yeast is used in brewing to make alcoholic drinks. Suggest why the yeast stops growing and dies before it has used up all the sugar in the fermentation solution.

3 Write a method to investigate which type of sugar provides the best substrate for anaerobic respiration. You have available three different sugars: granulated, castor, and demerara.

B4.2 How do we know about mitochondria and other cell structures?

A: Different types of microscope

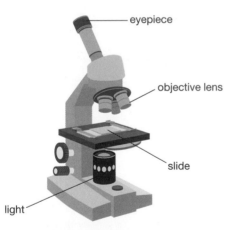

A typical light microscope.

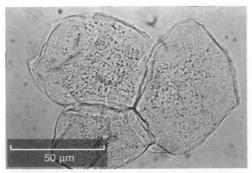

50 μm

Human cheek cells seen using a light microscope.

How do we know about mitochondria? They are far too small to be seen by the naked eye. The same is true for chloroplasts and other cell organelles.

We talk about these organelles when we explain where cellular respiration and photosynthesis take place. But how were they discovered? Sometimes, an explanation is only developed after a new piece of technology helps us to make new observations. Observations of cells and organelles only became possible after different types of microscope were invented.

The light microscope
Magnification

You may have used a **light microscope** in your lessons. We use microscopes to observe structures that are too small to be seen without **magnification**. Magnification makes objects appear bigger.

A light microscope uses visible light and two magnifying lenses (the objective lens and the eyepiece lens). The total magnification is calculated by multiplying the magnifying powers of the two lenses together.

For example, if:

- the magnification of the eyepiece lens is ×10
- the magnification of the objective lens is ×40

then the total magnification of the microscope is 10 × 40, which is ×400.

Resolution

Magnification is only useful if it is possible to see fine detail. A microscope that lets you see details that are very close together has high resolving power, or high **resolution**.

The resolution of light microscopes is limited by the nature of light. The smallest distance between points that can be seen using the best light microscopes is 0.0002 mm (0.2 μm). This limit means the maximum useful magnification provided by such microscopes is ×1500. Even if the magnification is higher, it is impossible to tell objects apart that are very close together. In fact, it is usually difficult to tell objects apart that are closer together than 1 μm, and very difficult to even see objects that are smaller than 1 μm in diameter.

Depth of field

The light microscope has to be focused exactly onto the structure being examined. Objects above and below appear blurred. This is because a light microscope has a very shallow **depth of field**.

Why is the electron microscope better?

The invention of the **electron microscope** enabled scientists to make much better observations of cells.

The electron microscope uses a beam of **electrons**, rather than a beam of light, to produce an image. The electron beams are focused by powerful magnets.

- A scanning electron microscope uses the electron beam to scan the surface of a specimen, and generate electrons from its surface. These electrons are collected by a detector and processed by a computer to form an image.

- A transmission electron microscope passes the beam of electrons through the specimen. Electrons passing through are captured and processed to form an image. When electrons pass through, that part of the image is light. When electrons are blocked, that part of the image is dark.

Magnification and resolution

Electrons are so small that an electron beam can pick out parts of an object that are very close together. This gives electron microscopes very high resolution. The high resolution means that higher magnification is possible. Some electron microscopes can magnify up to ×500 000.

Depth of field

Electron microscopes also have a much greater depth of field than light microscopes. This is particularly useful when trying to see objects in three dimensions.

Understanding cell structure

The development of the electron microscope made it much easier to observe cell organelles, such as mitochondria and chloroplasts, at much higher magnification. Mitochondria can be up to 3 μm in diameter and chloroplasts up to 5 μm, so they are just visible with the light microscope. However, the resolution of a light microscope is not good enough to understand their structure.

When looked at under the electron microscope, scientists saw the following features in each:

- outer membrane
- inner membrane, folded and arranged in different ways
- space between the outer and inner membranes
- space inside the inner membranes.

Scientists developed explanations about how the structures of mitochondria and chloroplasts relate to their roles in cellular respiration and photosynthesis. They did this by linking observations of the organelles' structure with ideas about the function of each organelle. When a membrane is folded, it increases its surface area. This suggested that the area of the membrane is important to the biochemical pathways in the organelle. When more than one membrane-bound space exists inside an organelle, it suggests that different chemical reactions take place in each.

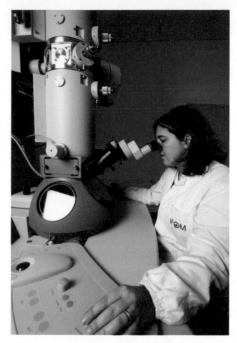

An electron microscope.

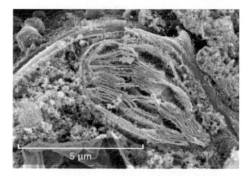

Scanning electron micrograph of a chloroplast inside a plant cell. The chloroplast has been coloured bright green by a computer, to make it easier to distinguish it from the other cell structures.

Questions

1 Write down the advantages of the electron microscope over the light microscope.

2 Explain why the internal structure of mitochondria is not clear under the light microscope.

3 Ribosomes are cell organelles used in making proteins. A ribosome is about 20 nm in diameter. Find out how big nanometres (nm) are, and then explain why ribosomes cannot be seen using a light microscope.

A: Cell division and growth

Find out about

- the role of the cell cycle in growth
- interphase and mitosis
- how cancer results from uncontrolled cell division

Key words

- ➤ unicellular
- ➤ multicellular
- ➤ cell cycle
- ➤ interphase
- ➤ mitosis
- ➤ tumour

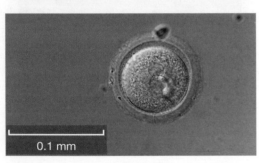

0.1 mm

Every human being begins life as a zygote – a single fertilised egg cell.

You began life as a single cell (a fertilised egg cell). But this cell divided many times to make all the cells in your body. By the time you were born, you were made of millions of cells. Your mass at birth was probably about 3 or 4 kg, but now it may be over 50 kg.

Some organisms are **unicellular** (made of just one cell) but most are **multicellular** (made of many cells). Growth of multicellular organisms involves an increase in the number of cells. All cells are made when a parent cell divides.

The cell cycle

Most cells go through different phases. The phases are part of the **cell cycle**. Looking at the cell cycle helps to understand how a cell divides. There are two main phases: **interphase** and **mitosis**.

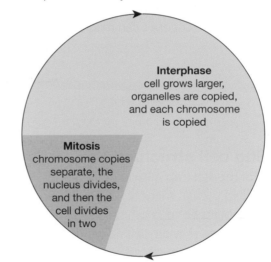

Interphase
cell grows larger, organelles are copied, and each chromosome is copied

Mitosis
chromosome copies separate, the nucleus divides, and then the cell divides in two

To enable organisms to grow, cells go through a cell cycle including a long interphase and a short mitosis phase.

Interphase

When new cells are made, they contain the same number of chromosomes as each other and the parent cell. They also contain the same organelles. In interphase, before dividing, the cell has to make copies of its organelles and its chromosomes. The cell also increases in size. In this phase, you cannot really see the chromosomes under the light microscope, because they unwind inside the nucleus so they can be copied. After the chromosomes have been copied, they wind up and become shorter and fatter.

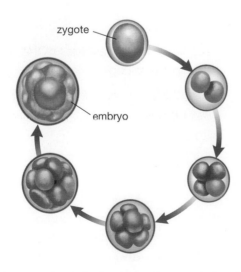

zygote

embryo

A zygote (top) divides many times to make all the cells in the human body.

Mitosis

During mitosis, the copies of chromosomes separate, the nucleus divides, and then the cell divides. When the cell divides, the membrane pinches inwards and two new cells are formed. In plant cells, a new cell wall also forms in between the two new cells. Each new cell has a complete set of chromosomes, and two new nuclei are formed. Each cell also receives a complete set of organelles.

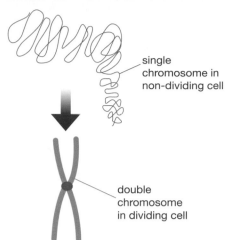

single chromosome in non-dividing cell

double chromosome in dividing cell

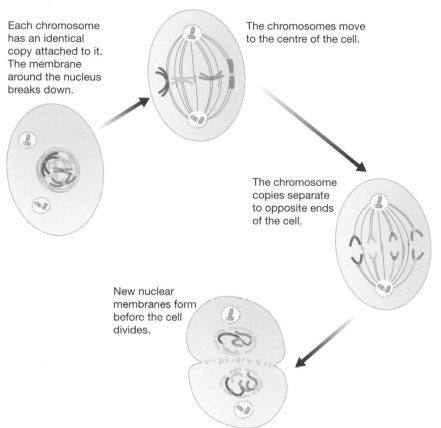

Each chromosome has an identical copy attached to it. The membrane around the nucleus breaks down.

The chromosomes move to the centre of the cell.

The chromosome copies separate to opposite ends of the cell.

New nuclear membranes form before the cell divides.

The stages of mitosis in an animal cell. For simplicity, only four chromosomes are shown.

During interphase, the cell is not dividing and the DNA is spread out. This enables it to be used to make proteins. It is also copied so that there are two copies of each chromosome ready for mitosis, when the cell divides into two.

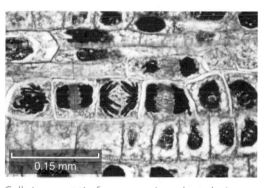

0.15 mm

Cells in a root tip from an onion plant during mitosis. The chromosomes have been stained red to make them easy to see using a light microscope. Different cells are in different stages of mitosis.

Cancer

Usually, genes tightly control the movement of a cell through the cell cycle. However, cancer is a non-communicable disease where changes (mutations) in these genes lead to uncontrolled cell division. The cell then divides many times by mitosis, eventually forming a ball of cells called a **tumour**.

If this happens in an important organ, it can lead to death. Sometimes, cells can also leave the tumour and move around the body. They continue to divide uncontrollably, causing tumours elsewhere. When this happens, we say the tumour is malignant.

Questions

1 What are the two main phases of the cell cycle?

2 Describe what happens during the two phases that lead to cell division.

3 Describe what causes a tumour to form.

B: Cell division and reproduction

Find out about

- the role of meiosis in making gametes
- the difference between meiosis and mitosis

Key words

➤ sexual reproduction
➤ fertilisation
➤ meiosis

All organisms reproduce – they produce offspring to ensure the continuation of the species. Humans, other animals, and flowering plants reproduce by **sexual reproduction**.

They produce gametes. The nuclei of two gametes join together during **fertilisation**, forming the first cell of a new living organism – the zygote.

- Males usually produce small gametes that move, such as human sperm cells produced in the testes.
- Females usually produce larger gametes, such as human egg cells produced in the ovaries.

Sexual reproduction in humans

Human body cells have 23 pairs of chromosomes, which makes 46 in total. But gametes only have one of each pair, or 23 single chromosomes. When the gametes join together in fertilisation, their nuclei fuse and the correct number of chromosomes (23 pairs, or 46 chromosomes in total) is restored in the zygote (23 chromosomes from the mother, and 23 from the father).

Sexual reproduction in flowering plants

Sexual reproduction in flowering plants happens when pollen grains from an anther are transferred to a stigma. The pollen grains can be transferred by the wind or by pollinators (usually insects). The pollen grains are not the male gametes, but they do contain the male gametes.

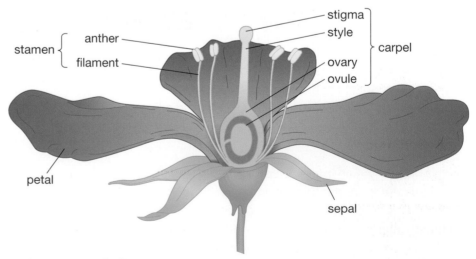

The structure of a flower.

Once it has reached the stigma, a pollen grain produces a pollen tube that carries the male gametes down the style to the ovary. There, female gametes are fertilised.

Making gametes – meiosis

Gametes are made by a type of cell division called **meiosis**. This process ensures that gametes end up with half the number of chromosomes of body cells.

The process of making gametes starts with normal body cells. These body cells go through interphase, and at the end of interphase there are two copies of each chromosome. The cells then divide twice – this is meiosis.

- The first division separates the copied chromosomes.
- The second division separates the copies of the 23 chromosomes.

This leaves half the number of chromosomes in each daughter cell, as seen in the diagram.

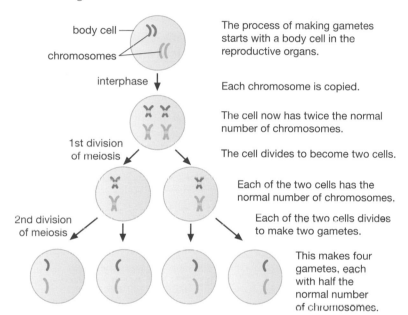

body cell
chromosomes

interphase

1st division of meiosis

2nd division of meiosis

The process of making gametes starts with a body cell in the reproductive organs.

Each chromosome is copied.

The cell now has twice the normal number of chromosomes.

The cell divides to become two cells.

Each of the two cells has the normal number of chromosomes.

Each of the two cells divides to make two gametes.

This makes four gametes, each with half the normal number of chromosomes.

The stages of making gametes. For simplicity, only four chromosomes (in two pairs) are shown at the start of the process.

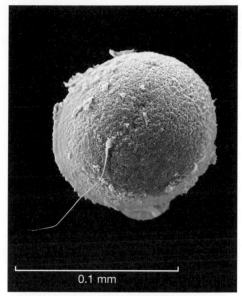

0.1 mm

This scanning electron micrograph shows a human sperm cell (artificially coloured blue by a computer) and a human egg cell (artificially coloured orange). These are human gametes, which fuse during fertilisation.

	Meiosis	Mitosis
Type of cells made	gametes (for sexual reproduction)	body cells (for growth and repair)
Number of cells at start of process	1	1
Number of cells at end of process	4	2
Number of rounds of cell division	2	1
Number of chromosomes in cell at the start of interphase	46 (23 pairs)	46 (23 pairs)
Number of chromosomes in each cell at end of process	23 single chromosomes (one from each pair)	46 (23 pairs)
Genetic variation in the cells made?	yes	no (unless an error has been made while copying DNA)

Questions

1 How many chromosomes does each person receive from their father and from their mother?

2 Why is it important that gametes have half the number of chromosomes of body cells?

3 Draw a flowchart to explain how gametes are made by meiosis.

C: Development

Key words

➤ development
➤ unspecialised
➤ embryonic stem cells
➤ specialised
➤ differentiation

You've grown a lot bigger since you were born. But you haven't just grown – you've developed too. **Development** is how an organism changes as it grows and matures. Your cells have been organised into different tissues and organs, and they have taken on specific jobs. Growth and development are both gradual processes, but development can involve some big changes.

Embryonic stem cells

All the cells in your body come from the original zygote, which is formed when the sperm cell fertilises the egg cell. The nucleus of each cell contains an exact copy of the original DNA.

During the first week of growth, the zygote divides by mitosis to form an embryo. Up to the eight-cell stage, all the cells are **unspecialised**, and are called **embryonic stem cells**. After this stage, they can become **specialised** into any sort of cell. This is known as **differentiation**.

Your body has more than 300 types of specialised cell, including muscle cells, nerve cells, and red blood cells, each of which work together with others as tissues. After about two months, the main organs have formed and the developing baby is called a fetus. Adults contain about 10^{14} cells.

Cell differentiation

Every cell has a copy of all of your genes. Different genes are switched on and off when different types of cell differentiate. When cells differentiate, they make the proteins needed to become a particular type of cell. Each gene is the instruction for a cell to make a different protein. By controlling what protein a cell makes, genes control how a cell develops.

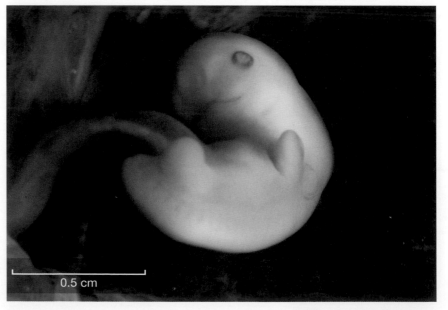

0.5 cm

This is a human embryo approximately 6 weeks after fertilisation. Cells in the body have already started to make arms. Cells at the end of the arms will make fingers. This happens because of the difference in the concentrations of chemical signals in each region of the embryo.

Gene switching

Each gene controls the manufacture of one type of protein. So, in any organism, there are as many genes as there are different types of protein. In humans there are 20 000–25 000 genes. Not all these genes are active in every cell. As cells grow and specialise some genes switch off.

In a hair cell, the genes for the enzymes that make keratin will be switched on. But the genes for the enzymes that make amylase will be switched off:

hair cell genes switched on ⟶ *enzymes for making keratin* ⟶ *hair grows*

In a salivary gland cell, those genes for making amylase will be switched on:

salivary gland cell genes switched on ⟶ *amylase secreted* ⟶ *starch digested*

Gene switching in embryos

An early embryo is made entirely of embryonic stem cells. These cells are unspecialised up to the eight-cell stage. All the genes in these cells are switched on. As the embryo develops into a fetus, cells specialise. Different genes switch off in different cells. As we grow, the position and type of cells must be controlled, so each tissue and organ develops in the right place.

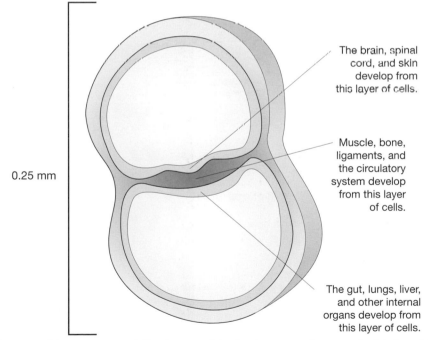

0.25 mm

The brain, spinal cord, and skin develop from this layer of cells.

Muscle, bone, ligaments, and the circulatory system develop from this layer of cells.

The gut, lungs, liver, and other internal organs develop from this layer of cells.

It is possible to map specialised parts of the body onto a diagram of a 14-day-old embryo. Here you can see which groups of cells in the embryo will develop into future tissues and organs.

Some genes, such as those that produce the enzymes needed for respiration, stay switched on in every cell.

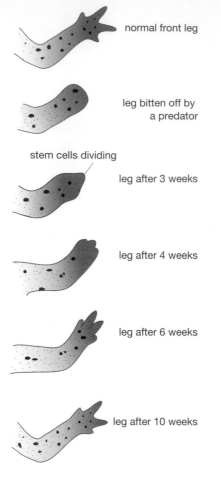

normal front leg

leg bitten off by a predator

stem cells dividing

leg after 3 weeks

leg after 4 weeks

leg after 6 weeks

leg after 10 weeks

If a newt's limb is bitten off by a predator, it can grow a replacement. Most animals can only make small repairs to their body.

Questions

1 Name the type of cell in a plant that can divide to make new cells.

2 Why is it important that all living things have cells that can divide?

3 Explain why a newt can regrow a leg, but a human cannot.

4 Explain why and how unspecialised cells become differentiated.

Adult stem cells

In newts, stem cells stay unspecialised throughout their lives. This means that newts can grow new body parts if they need to. **Adult stem cells** in humans are less useful. They have already started to specialise. For example, the stem cells in your skin can only develop into skin cells.

Other parts of the body also need a constant supply of new cells. For example, bone marrow contains stem cells to make new blood cells.

Plant stem cells

Flowering plants continue to grow throughout their lives.

● Their stems grow taller.

● Their roots grow longer.

● To hold themselves upright, most stems increase in girth (circumference) or have some other means of support.

Plants increase in length by making new cells through mitosis at the tips of both shoots and roots. They also have rings of dividing cells in their stems and roots to increase their girth. These dividing cells are called **meristem** cells.

Because plants have meristem cells, it means they can grow new xylem and phloem, and they can regrow whole organs, such as leaves, if they are damaged.

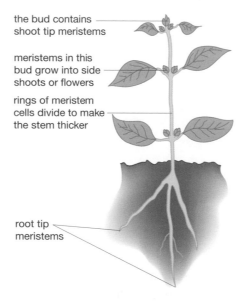

the bud contains shoot tip meristems

meristems in this bud grow into side shoots or flowers

rings of meristem cells divide to make the stem thicker

root tip meristems

Meristem cells divide to make stems and roots longer, and make the stem thicker.

This giant Sequoia tree is over 2000 years old, 83 m tall, and 26 m in girth. Parts of it have been growing all that time. It is now the largest tree on Earth.

B4.4 Should we use stem cells to treat damage and disease?

A: Stem cell treatments

Because stem cells are able to differentiate into different types of cell, we could use them in medicine to help treat damage and disease. **Stem cell treatment** is already in use in some countries, and many more applications of stem cells are currently being trialled. However, the number of treatments currently approved is still very small.

Benefits of stem cell treatments

Blood disorders

For over 30 years, blood disorders, such as leukaemia, have been treated with bone marrow transplants. Bone marrow contains stem cells that divide and differentiate to make eight different kinds of blood cell. Leukaemia is a kind of cancer where the body makes too many white blood cells. Treatment involves killing the patient's own bone marrow cells with radiation. New bone marrow from a donor can make healthy blood.

Spinal cord and brain damage

Stem cells may have the potential to cure people who are paralysed through injuries to their spinal cord. Injecting stem cells into the damaged area could lead to regrowth of nerve cells. In the same way, it may be possible to treat brain damage, such as that caused by Parkinson's or Alzheimer's disease. Research is still at clinical trial stage.

Diabetes

If stem cells will differentiate into insulin-secreting cells in the pancreas, it may be possible to cure some types of diabetes. Research is still at the clinical trial stage.

Risks of stem cell treatments

Some of the risks of stem cell treatments include:

- infections – from operations or from infected stem cells
- tumours – due to uncontrolled cell division
- organ damage – when stem cells implanted into damaged tissue migrate to healthy areas they can cause damage to these areas
- immune rejection
- **graft-versus-host disease**.

Immune rejection and graft-versus-host disease only happen when the stem cells come from a person other than the patient. Immune rejection happens when the patient's white blood cells recognise the donor's stem cells as non-self, and attack them. Graft-versus-host disease can occur when bone marrow stem cells have been donated, and the new bone marrow makes white blood cells. The white blood cells from the new bone marrow recognise the patient's body cells as non-self, and attack them.

Find out about

- potential uses of stem cells in medicine
- the benefits, risks, and ethical issues

Key words

➤ stem cell treatment
➤ graft-versus-host disease

Stem cell treatments could offer a potential new treatment for people with Alzheimer's disease.

Synoptic link

You can learn more about diabetes in B5.6A *Controlling blood sugar levels*.

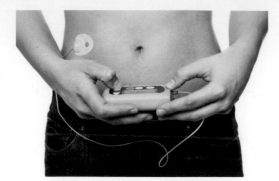

Some people with diabetes have to give themselves regular injections of a hormone called insulin. Stem cell treatments could help their bodies to produce its own insulin.

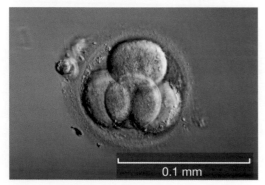

0.1 mm

Very early human embryos, such as this one at the four-cell stage, can provide embryonic stem cells.

The Human Tissue Authority helps to regulate the use of stem cells in the UK.

Where would we get stem cells to be used in treatments?

If we are going to use stem cells to treat medical conditions, we need a source of stem cells.

Adult stem cells

Each of us has stem cells in our tissues. These are adult stem cells. But adult stem cells can only differentiate into a small number of cell types. Because of this, their uses are limited.

For example, we all have some stem cells in our skin, eyes, blood, and bone marrow. A fault in one eye could be treated by stem cells from the other eye. But adult stem cells from the eyes would not be able to differentiate into the types of cells needed to treat a blood disorder or diabetes.

Embryonic stem cells

Embryonic stem cells can differentiate into any type of cell. Because of this, they may be more useful for medicine and research.

Embryos left over from fertility treatments are a source of embryonic stem cells. Only very early embryos, made up of a few cells, can be used. These cells are completely unspecialised and can be kept in the laboratory to produce more stem cells, and to produce differentiated cells. However, many people think it is wrong to use embryos to harvest stem cells.

There may be an alternative. With special treatment, specialised body cells can sometimes be made to behave like embryonic stem cells. This may solve many of the ethical issues with using embryonic stem cells. However, stem cells made this way are currently only authorised for use in research, not in treatments.

Regulation

Use of stem cells, particularly embryonic stem cells, is subject to **regulation**. The regulation is enforced by bodies that are independent of politicians and independent of the medical professionals involved. These bodies must consider:

● the balance between benefit and risk

● the ethical issues.

In the UK, this is done by the Human Fertilisation and Embryology Authority and the Human Tissue Authority.

When assessing the risks of a scientific technology, members of the public may not be informed enough about the science, its benefits, and potential **hazards** to make their own judgement. Because of this, governments try to encourage public trust in a regulatory body, which is informed, and which should be able to make decisions on their behalf. Although it is important that the public is safe, some scientists think that stem cell treatments may be over-regulated.

If someone is very ill, they (or their family) may want a new stem cell therapy to be used. For this reason, some countries are allowing stem cell therapy to be used:

- for very serious, potentially fatal diseases where there is no other therapy available
- even though the stem cell therapy has not been fully tested and legally approved.

This kind of approach has started to influence regulators' thinking. The European Medicines Agency thinks that regulators are too scared about risks, and do not consider both potential benefits and risks when approving a medical treatment. They say that regulators should:

- include the patient's view on the level of acceptable risk, which may be different to the regulator's view
- recognise that no medicine is 'risk-free'
- work out ways to combine patients' opinions with hard numerical trials data when coming to a conclusion about a medicine.

This should make public regulation of risk less controversial, and allow potentially beneficial medical treatments to reach the patient much earlier.

Ethical decisions about stem cells

The use of stem cells is particularly controversial because of the use of human embryos. People tend to hold a position most closely aligned to one of those below, each of which balances 'what they feel is right' against 'consequences'.

Jayshree
An embryo is a human being. We would be horrified if someone was murdered to provide spare parts for someone else. Using a human embryo in this way is exactly the same. No positive consequences can justify it.

Connor
I don't think an embryo is just a ball of cells, but I'm not sure it is yet a human being. I can see the positive benefits of stem cell therapy.

Peter
I don't think that a ball of eight cells can yet be considered to be a human. So long as the risks are minimised, the positive benefits of stem cell therapy are too important to give up.

Connor's view is the basis of most UK law about using embryonic stem cells in research and medicine. It might be called the 'no ... unless' view. This approach has the following implications:

- The researcher must look for an alternative to embryonic stem cells, only using them when there is absolutely no alternative.
- The research should be about developing treatment for a significant medical need, such as a serious disease or crippling injury.
- Research should only be allowed up to a certain point in the embryo's development.

Questions

1 What is the difference between stem cells and differentiated cells, and why is the difference important?

2 What are the advantages of using stem cells from your own body to treat an illness?

3 Why is stem cell treatment with embryonic stem cells controversial?

Science explanations

B4 Using food and controlling growth

In this chapter you learnt about the importance of cellular respiration to all living cells, the importance of cell division in growth and reproduction, and the importance of cell differentiation in development. You also learnt about applications of science, such as in stem cells, which offer tremendous benefits but also come with risks and ethical questions.

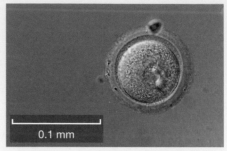

We all began life as a single cell. Mitosis and differentiation created the billions of different cells in your body. Cellular respiration keeps those cells alive.

You should know:

- the differences between aerobic and anaerobic respiration, including the conditions under which they occur, their inputs and outputs, and their relative yields of ATP

- why cellular respiration occurs continuously in living cells

- about the role of mitochondria in cellular respiration in animal and plant cells

- why cellular respiration is described as an exothermic reaction

- how to calculate the rate of cellular respiration

- how electron microscopy has increased our understanding of sub-cellular structures

- about the role of the cell cycle in growth, including interphase and mitosis

- that cancer results from uncontrolled cell division due to changes in genes

- about the role of mitosis in halving the chromosome number to form gametes

- about the function of stem cells in embryos and adults

- that cells differentiate to become specialised, and that this involves switching genes off and on

- about the function of meristems in plants

- about the potential benefits, risks, and ethical issues associated with the use of stem cells in medicine.

Ideas about Science

Collecting enough data to enable an explanation to be developed sometimes relies on technological developments. New technology enables new observations to be made, and new explanations to be developed. The invention of the electron microscope enabled scientists to image the inside of cells, and the inside of organelles, in more detail than ever before. These observations helped us to explain the roles of mitochondria and chloroplasts, and the processes of cellular respiration and photosynthesis, more clearly.

You should be able to describe and explain examples of scientific explanations that were modified when new evidence became available.

Applications of science can help us to improve our quality of life. For example, using plant hormones to control plant growth and ripening can help us feed the population, while stem cell treatments could be used to treat damage and disease. But everything we do carries risk, and new technologies can introduce new risks. Some applications of scientific knowledge have ethical implications, for example, when stem cells are sourced from embryos.

Governments and public bodies act to regulate applications of science and technology, but their decisions can be controversial. They make decisions based on an analysis of benefits, risks, and ethical issues.

You should be able to describe ethical issues associated with an application of science, and summarise different views that may be held.

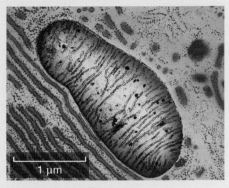

The invention of the electron microscope enabled scientists to see mitochondria in incredible detail. This helped us to explain their role in cellular respiration.

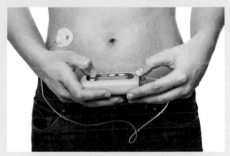

Stem cell treatment could help people with diabetes to live without insulin injections. But not everybody agrees that using stem cells should be allowed.

B4 Review questions

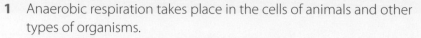

1 Anaerobic respiration takes place in the cells of animals and other types of organisms.

  a Describe conditions where anaerobic respiration is an advantage to:

    i an animal

    ii another type of organism.

  b Give an example of how people use anaerobic respiration in microorganisms to make a useful product.

2 ATP is an essential molecule in cells.

  a Write down the names of two processes that make ATP in cells.

  b Explain why the uptake of nitrate ions into plant root hair cells requires ATP, but the movement of carbon dioxide molecules through open stomata does not.

3 Ben investigates how the concentration of glucose affects the rate of anaerobic respiration in yeast. He sets up the following apparatus.

He places the conical flask into a water bath for 4 minutes. He writes down the volume of gas in the syringe each minute.

He repeats the experiment with glucose solution at a concentration of 0.5 g/dm³, and then again at 1.0 g/dm³.

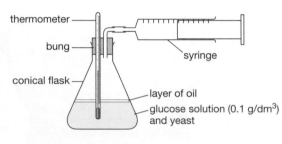

  a Why is a layer of oil added on top of the glucose solution and yeast?

  b Why is a bung used to seal the top of the conical flask?

  c Why is the flask placed in a water bath during the 4-minute data collection period?

  d Bubbles of gas form in the glucose solution. What are the bubbles made of? Choose the correct answer from the following list.

    oxygen gas   water vapour

    carbon dioxide gas   air

  e The results from one conical flask are shown on the graph. The concentration of the glucose solution in this flask was 0.1 g/dm³.

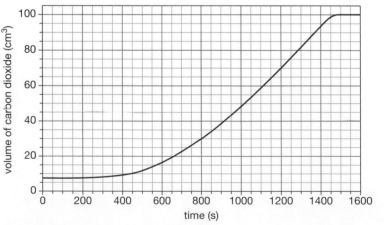

What is the average rate of cellular respiration between 400 s and 800 s in this flask?

f Predict how the graph could look different for the flask containing
glucose solution at 1.0 g/dm³, and explain why you have made
this prediction.

4 Four people try to explain the link between meiosis and fertilisation.

Jo
Fertilisation causes
the zygote to have
the full chromosome
number.

Lee
Fertilisation
avoids the
chromosomes
mixing together
inside the zygote
nucleus.

Sue
Since the number
of chromosomes
in gametes is
halved, twice as
many gametes can
be produced.

Ray
Meiosis produces
gametes with 23
chromosomes each.

Which two people's ideas together give the best explanation of the
link between meiosis and fertilisation?

5 A patient is diagnosed with cancer.

a Describe how changes in genes in a cell can lead to cancer.

b List two things that could cause these changes in genes.

c The patient's tumour is malignant. What does this mean?

d Suggest why the patient's immune system does not attack
the tumour.

6 A doctor is treating a patient who has leukaemia.

a Describe how stem cells from a donor could be used to help treat
this patient.

b One risk of this stem cell treatment is graft-versus-host disease.
Explain what this is.

c Write down three other risks of this stem cell treatment.

d Human embryos could provide a source of stem cells for this
treatment. What are the concerns that some people have about
using stem cells from human embryos?

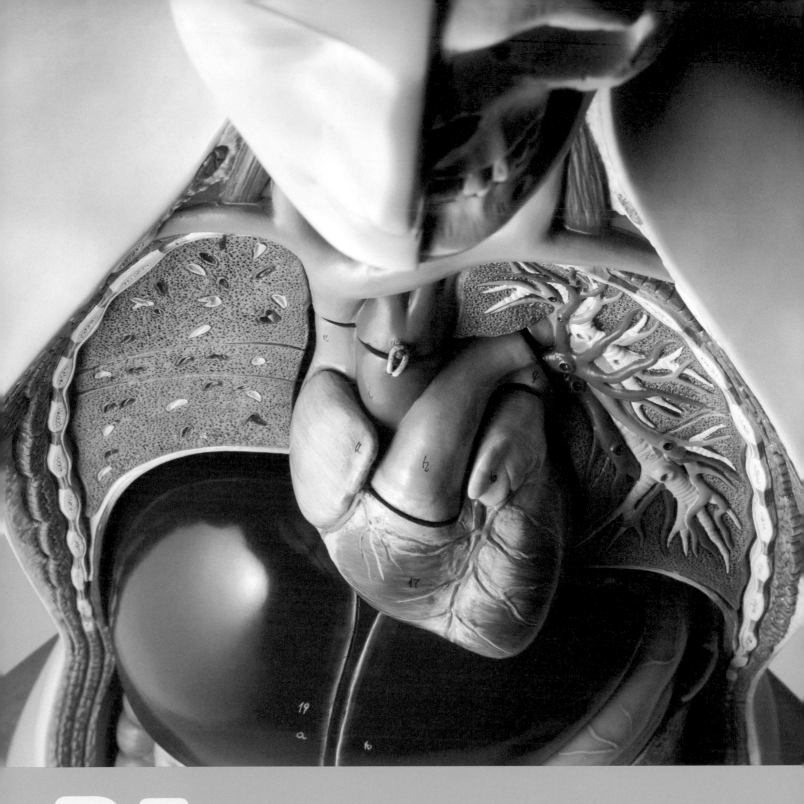

B5 The human body – staying alive

Why study the human body?

The human body is made of trillions of cells, organised into tissues, organs, and organ systems. The ways in which these systems work together is what supports the life processes of each cell and of the body as a whole. Systems in the body enable you to respond to changes in your environment, and – ultimately – to stay alive.

What you already know

- The cells of multicellular organisms are organised into tissues, organs, and organ systems.

- Muscles are tissues that can contract, and the contraction of muscles (often working in pairs) is what enables us to move.

- The tissues and organs of the human digestive system are adapted to move food through the body, break it down, and absorb substances from it.

- The structures of the human gaseous exchange system are adapted to move air into and out of the lungs.

- The lungs are adapted to absorb oxygen from the air and get rid of carbon dioxide from the body.

- The human circulatory system transports substances around the human body.

- The human body has sense organs that collect information about the world around us.

- People with diabetes cannot control their blood sugar levels.

- Factors such as obesity increase the risk of developing diabetes, and diabetes is itself a risk factor for cardiovascular diseases.

- Stem cell treatments offer the potential to cure some diseases, but there are risks and ethical issues associated with using them.

The Science

The substances essential for chemical reactions in cells are transported into, out of, and around the human body by the circulatory system, the gaseous exchange system, the digestive system, and the excretory system all working together. Some tissues are highly adapted for efficient exchange of substances between the body and its surroundings.

Detecting and responding to changes is essential for survival, and depends on the nervous system and hormones. This enables you to maintain a constant, ideal environment inside your body. Hormones are also essential for reproduction.

Sometimes body systems can be damaged by accidents or disease. Scientists use their understanding of the body to help develop treatments.

Ideas about Science

Understanding the role of hormones in reproduction has enabled us to develop contraceptives and also treatments for infertility. Understanding of the nervous system has led to treatments for damage to sense organs and neurons. These applications of science could change lives for the better, but what are the issues with using them?

A: Supply routes in humans

Find out about

- some of the substances transported into and out of the human body
- the relationship between the circulatory system, the gaseous exchange system, the digestive system, and the excretory system
- how substances move between systems

Key words

➤ circulatory system
➤ gaseous exchange system
➤ digestive system
➤ excretory system
➤ urea

Inside every cell in your body thousands of chemical reactions are happening every second. These reactions are keeping you alive. They need a constant supply of water, oxygen, and substances from food. Waste products have to be continually removed. If they weren't, they would build up to toxic concentrations and your body would poison itself.

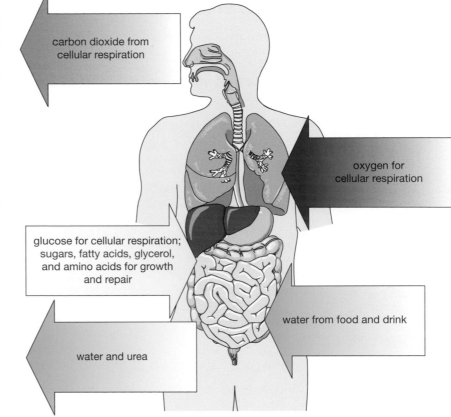

carbon dioxide from cellular respiration

oxygen for cellular respiration

glucose for cellular respiration; sugars, fatty acids, glycerol, and amino acids for growth and repair

water from food and drink

water and urea

Some of the inputs and outputs of the human body.

How are all of these different substances transported into, out of, and around the human body?

Cells to systems

An adult human is made of approximately 37.2 trillion cells (that's 37.2 million million, or 3.72×10^{13}). Keeping all of those cells alive is not an easy task. The way the body is organised is very important. Groups of identical cells work together as tissues. Tissues work together to form organs. Organs work together as organ systems, and organ systems work together to keep cells alive.

Every cell in your body, from your head down to your toes, is kept alive by organ systems working together.

Systems working together

There are a number of important organ systems in your body, including:

- the **circulatory system** – including the heart, blood vessels, and blood

- the **gaseous exchange system** – including the nose and mouth, trachea (windpipe), and lungs

- the **digestive system** – including the mouth, oesophagus, stomach, liver, pancreas, intestines, rectum, and anus

- the **excretory system** – including the lungs, kidneys, bladder, and skin.

The circulatory system transports substances around the body. The gaseous exchange system takes in oxygen from the air we breathe in (and adds carbon dioxide to the air we breathe out). The digestive system breaks down the biomass in the food we eat. The excretory system gets rid of excess water, **urea** (in urine), salts, and carbon dioxide.

All of these systems work together to move substances into, around, and out of the body.

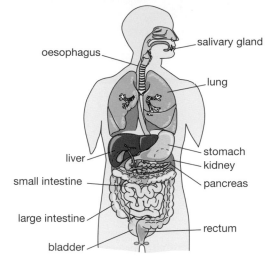

Important organs in the human body. You can look up each organ in the glossary to find out what it does.

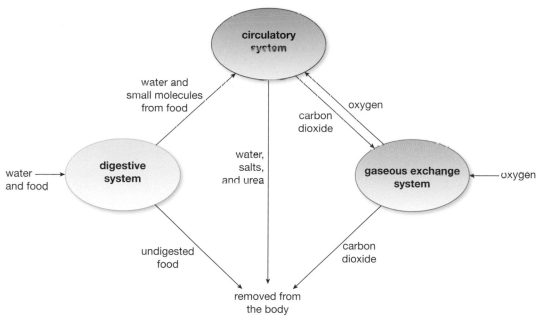

Substances move between different systems in your body.

How do substances move between systems?

Substances have to move from one organ system to another. Usually, the cells and fluids of different systems are separated by partially permeable membranes. Substances move through these membranes by diffusion, osmosis, and active transport.

Gases and small molecules such as urea move through partially permeable membranes by diffusion. Water always moves by osmosis. Larger molecules such as lactic acid, urea, sugars, fatty acids, glycerol, and amino acids are moved through partially permeable membranes by active transport.

Synoptic link

You can learn more about diffusion, osmosis, and active transport in B3.2A *Moving molecules*.

Moving between the gaseous exchange system and the circulatory system

Your lungs are divided up into millions of tiny sacs called **alveoli** (singular alveolus). Each one fills with air when you breathe in. The air is exhaled when you breathe out.

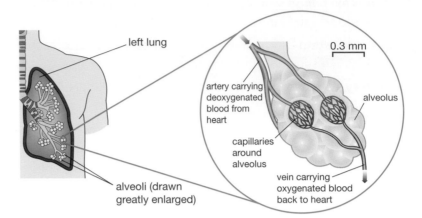

Gaseous exchange in the lungs. Each alveolus is wrapped in a network of blood vessels called capillaries. The walls of the alveoli and the capillaries are very thin, so gases can diffuse through quickly.

Oxygen diffuses from the air in each alveolus into red blood cells in capillaries (blood vessels). Carbon dioxide diffuses from the blood plasma into the air in the alveoli.

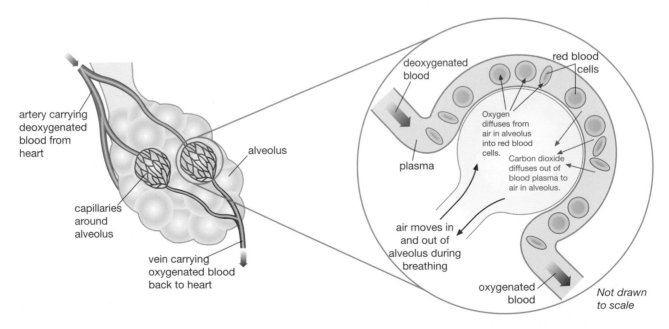

A simple representation of the movement of gases between the air in an alveolus and the blood.

Moving between the digestive system and the circulatory system

Water and molecules from food are absorbed by the lining of the small intestine. The inside wall of the small intestine has many finger-like projections called **villi** (singular villus) on its surface. Each villus contains a network of blood capillaries. Substances from the intestine move through the thin wall of the villus into the capillaries.

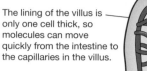
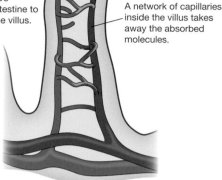

The lining of the villus is only one cell thick, so molecules can move quickly from the intestine to the capillaries in the villus.

A network of capillaries inside the villus takes away the absorbed molecules.

0.5 mm

The structure of a villus enables water and molecules from food to be absorbed into the bloodstream quickly.

Water moves between the digestive system and the blood plasma by osmosis. Molecules from digested food (including sugars, fatty acids, glycerol, and amino acids) are too large to diffuse through partially permeable membranes – they are absorbed by active transport.

Supplying cells and removing waste

The circulatory system carries oxygen, water, and molecules from food around your body to supply your cells. Oxygen diffuses from the blood plasma into cells. Water moves into cells by osmosis. Molecules from food are moved into cells by active transport.

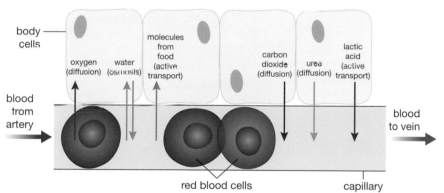

body cells

oxygen (diffusion) water (osmosis) molecules from food (active transport) carbon dioxide (diffusion) urea (diffusion) lactic acid (active transport)

blood from artery

blood to vein

red blood cells   capillary

A simplified diagram showing the exchange of substances between the blood and body cells.

Waste products from cellular respiration include carbon dioxide and lactic acid. Carbon dioxide diffuses out of cells into the blood plasma. Lactic acid is moved by active transport.

Urea is a waste product made during the breakdown of other substances in the liver. Molecules of urea are small enough to move through partially permeable membranes by diffusion. Urea diffuses out of liver cells into the blood plasma. When the blood reaches the kidney, urea diffuses out of the blood plasma into collecting tubes in the kidney, to form urine. The urine is excreted through the bladder when you go to the toilet.

Questions

1 Name two substances that are transported around the body in the blood.

2 Write down two substances that are absorbed from the gut into the blood plasma by active transport.

3 Describe the relationship between the circulatory system and the gaseous exchange system.

B: Moving blood around

Find out about

- how the structure of the human heart is adapted to its function
- how the structures of arteries, capillaries, and veins are adapted to their functions
- red blood cells and blood plasma

Key words

➤ atria
➤ ventricles
➤ deoxygenated blood
➤ oxygenated blood
➤ valve

Substances are transported around your body in the blood. Your blood is pumped through your blood vessels by your heart.

A double circulation

The blood vessels form a continuous system around the human body. They carry blood through the heart twice on each complete trip around the body. This is because humans have a double circulation system.

Blood goes:

- from the heart to the lungs to pick up oxygen (and get rid of carbon dioxide)
- back to the heart for a pressure boost
- around the body to deliver oxygen (and transport other substances)
- back to the heart again.

The human heart

The heart is the pump that makes blood circulate. The heart is made of four muscular chambers. The upper chambers receive blood, and are called **atria** (a single chamber is called an atrium). The lower two chambers have thick muscular walls to pump blood out of the heart, and are called **ventricles**. The heart muscles contract automatically, and the rate varies with the body's level of activity and stress. When you are sitting down, your heart contracts at about 70 beats per minute.

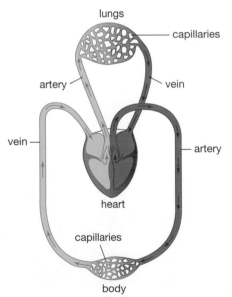

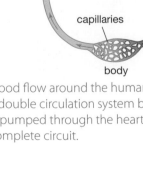

Blood flow around the human body. We have a double circulation system because blood is pumped through the heart twice in each complete circuit.

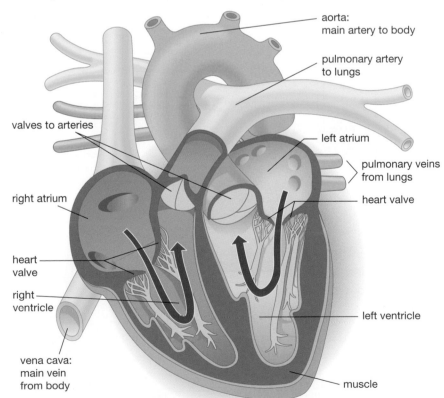

Diagram of a human heart: 'left' and 'right' refer to the way your heart sits in your body. The blood vessels and heart chambers that have been coloured blue carry blood that is short of oxygen. Those coloured red carry blood containing plenty of oxygen.

The right atrium receives blood from the body. Most of the oxygen in this blood has been used by cells – it is **deoxygenated blood**. The right atrium pumps the blood into the right ventricle. The right ventricle pumps the blood to the lungs to pick up oxygen. **Oxygenated blood** returns to the heart and is received in the left atrium. The blood is pumped into the left ventricle. The left ventricle then pumps oxygenated blood to the body.

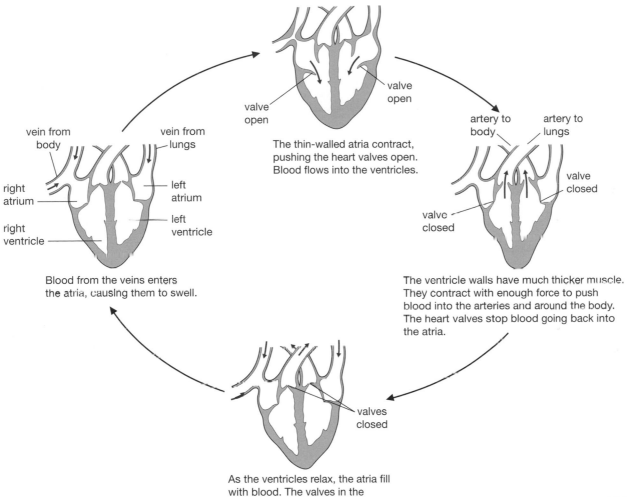

valve open

valve open

The thin-walled atria contract, pushing the heart valves open. Blood flows into the ventricles.

artery to body

artery to lungs

valve closed

valve closed

The ventricle walls have much thicker muscle. They contract with enough force to push blood into the arteries and around the body. The heart valves stop blood going back into the atria.

vein from body

vein from lungs

right atrium

left atrium

right ventricle

left ventricle

Blood from the veins enters the atria, causing them to swell.

valves closed

As the ventricles relax, the atria fill with blood. The valves in the arteries stop backflow into the heart.

The sequence of events in a heartbeat. Both sides of the heart contract at the same time.

A one-way system

When the ventricles contract they push blood out of the heart. But what stops the blood from going back up into the atria? This is the job of the heart **valves**. They act like one-way doors to keep the blood flowing in one direction. There are two sets of valves in the heart:

- between each atrium and ventricle – these valves stop blood flowing backwards from the ventricles into the atria
- between the ventricles and the arteries leaving the heart – these valves stop blood flowing backwards from the arteries into the ventricles.

The comforting 'lub-dub' sound of a heartbeat is the sound of heart valves snapping shut. The 'lub' sound is caused by the valves between the atria and ventricles shutting. The 'dub' sound is made as the valves between the ventricles and the arteries close.

Blood vessels

You have three different types of blood vessels: **arteries**, **capillaries**, and **veins**. Each type is adapted to its function in different ways.

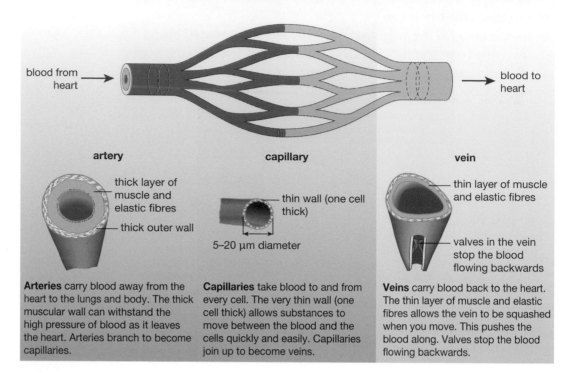

artery

- thick layer of muscle and elastic fibres
- thick outer wall

capillary

- thin wall (one cell thick)

5–20 µm diameter

vein

- thin layer of muscle and elastic fibres
- valves in the vein stop the blood flowing backwards

blood from heart →

blood to heart →

Arteries carry blood away from the heart to the lungs and body. The thick muscular wall can withstand the high pressure of blood as it leaves the heart. Arteries branch to become capillaries.

Capillaries take blood to and from every cell. The very thin wall (one cell thick) allows substances to move between the blood and the cells quickly and easily. Capillaries join up to become veins.

Veins carry blood back to the heart. The thin layer of muscle and elastic fibres allows the vein to be squashed when you move. This pushes the blood along. Valves stop the blood flowing backwards.

- valve only lets blood flow towards the heart
- leg muscles squeeze blood upwards
- valve closes to stop blood going backwards

Blood in veins is at a lower pressure than blood in arteries. When you move around, your body muscles help push the blood along veins. Valves stop the blood flowing backwards.

Exchanging substances

On average you have six litres of blood in your body. It is all pumped through the heart three times each minute. However, the blood spends most of that time in capillaries. The lungs and other body organs contain beds or networks of capillaries. Capillaries are where substances from the blood and the body's cells are exchanged. The structure of capillaries means they are well adapted for this function.

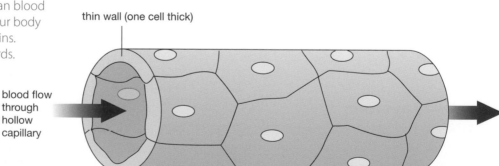

thin wall (one cell thick)

blood flow through hollow capillary

The thin wall of a capillary allows substances to be exchanged quickly and easily with cells.

Blood

A sample of blood looks completely red. A closer look shows that it consists of cells floating in pale yellow liquid. Some of the cells are white blood cells, and some are cell fragments (platelets). You learnt about these in B2. The rest of the cells are **red blood cells**.

Red blood cells

Red blood cells are the most obvious blood component because of their number and colour. The cells are adapted to their function of transport. They are packed with the protein **haemoglobin**. Haemoglobin binds with oxygen as blood passes through the lungs. The oxygen is released from haemoglobin as blood circulates through the tissues of the body.

Red blood cells have no nucleus, which allows more space for haemoglobin. If the haemoglobin circulated freely in the plasma instead, then the blood would be too thick to flow properly.

The **biconcave** shape gives the cells a large surface area through which oxygen can diffuse. This shape also gives cells flexibility to squeeze through tiny capillaries.

Plasma

The liquid part of blood is called **plasma**. Plasma is mainly water. Oxygen is carried by red blood cells, but all other substances transported by the blood are carried by the plasma. The plasma acts as a solvent. It gives blood its bulk and also helps to distribute heat around the body.

Key words

➤ artery
➤ capillary
➤ vein
➤ red blood cell
➤ haemoglobin
➤ biconcave
➤ plasma

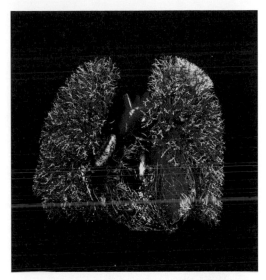

This resin cast shows all the capillaries of an adult human's lungs.

This is what blood cells look like using an electron microscope. In this image red blood cells have been coloured red by a computer, white blood cells coloured yellow, and platelets coloured pink. Each 1 mm³ of blood contains approximately 5 million red blood cells, 250 000 platelets, and 7000 white blood cells.

Questions

1 Explain what is meant by a double circulatory system.

2 Choose the best option from the list to complete each of the following sentences about veins.
   blue
   deoxygenated
   high pressure
   oxygenated
   thick walls
   thin walls
   valves
   a Veins carry .... blood from the body cells to the heart.
   b Veins have .... to stop blood from flowing backwards.
   c Veins carry .... blood from the lungs to the heart.

3 Explain how the structures of arteries, capillaries, and veins are adapted to their function.

4 Different parts of the heart have muscular walls of different thickness. Explain the difference in thickness between:
   a atria and ventricles
   b the right and left ventricles.

C: The importance of transport and exchange surfaces

Find out about

- why multicellular organisms need a transport system
- exchange surfaces
- surface area:volume ratio

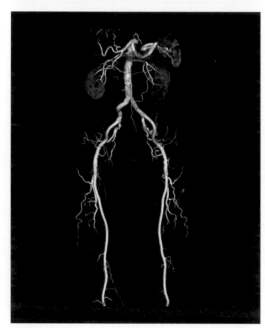

This is not a bolt of lightning. It's a computed tomography (CT) scan showing the arteries in an adult human's lower abdomen and legs. Your feet are a long way from your lungs, but arteries carry oxygenated blood there very quickly.

Keeping all of your cells alive depends on each one receiving a constant supply of oxygen, glucose, and water, and removing waste. But many of your cells are buried deep inside your body. They're a long way from the air outside, with its plentiful supply of oxygen. Some of them are a long way from the air in your lungs and from the food in your digestive system. Keeping these cells supplied, and removing their waste, is a problem faced by all multicellular organisms.

The need for a transport system

Life is simpler when you're an amoeba. For organisms that are made of a single cell, supplying the cell with essential substances and removing its waste are easy. Substances move directly into and out of the organism through the cell surface membrane.

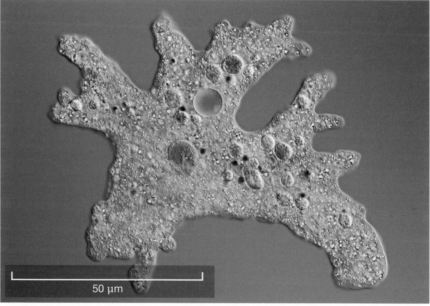

50 μm

An amoeba is an organism made of just one cell. Substances move directly into and out of the organism through the cell surface membrane.

Humans, other animals, and plants are multicellular organisms. It would take too long for oxygen and other essential molecules to diffuse from outside our bodies to all of the trillions of cells inside. The distance is too great. The evolution of complex multicellular organisms depended on the evolution of transport systems. These systems carry essential substances and waste around the body quickly. Humans and other animals have the heart and blood vessels of the circulatory system. Plants have xylem and phloem.

All human organs and tissues contain capillary beds, so the distance that substances have to diffuse between the blood and cells is very small.

Exchange surfaces

Multicellular organisms have **exchange surfaces** where substances are absorbed. For example, the lining of the small intestine is an exchange surface. The villi greatly increase the **surface area** for absorption of water and molecules from food. The alveoli in the lungs and root hair cells in plants do the same thing – they are also exchange surfaces.

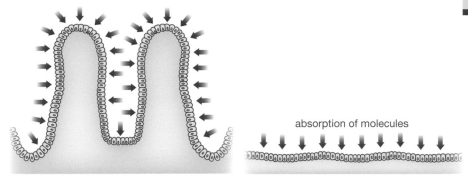

surface with villi

absorption of molecules

surface without villi

Compare these two surfaces. The surface with villi has a much greater surface area through which to absorb molecules than the surface without villi.

Surface area:volume ratio

The movement of substances into organisms of different sizes can be modelled using cubes. If you placed a cube of agar jelly into coloured water, you would see the coloured dye diffusing into the cube. It would diffuse from the surface of the cube towards the middle. Eventually the cube would completely change colour.

Diffusion of a substance from the surface of a three-dimensional shape to its centre is more efficient when:

- the distance that the substance has to diffuse is shorter
- the surface area of the shape is larger relative to its volume – we call this the **surface area:volume ratio**.

Key words

➤ exchange surface
➤ surface area
➤ surface area:volume ratio

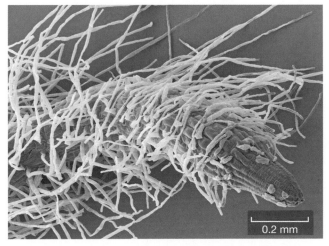

An electron micrograph showing root hairs (artificially coloured white) on a plant root (orange). The root hairs greatly increase the surface area for absorption of water and mineral ions.

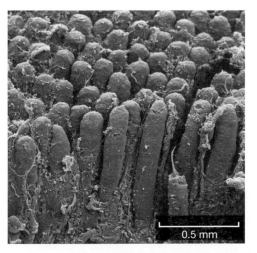

An electron micrograph of the wall of the small intestine. The finger-like projections are villi. They greatly increase the surface area for absorption.

Imagine three experiments using agar cubes, as follows:

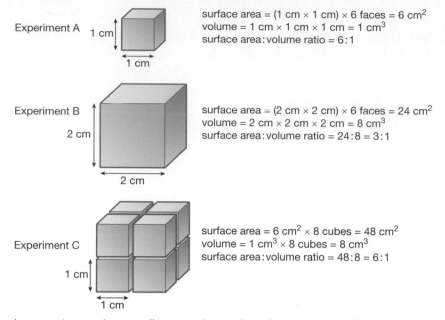

Experiment A
1 cm
1 cm

surface area = (1 cm × 1 cm) × 6 faces = 6 cm²
volume = 1 cm × 1 cm × 1 cm = 1 cm³
surface area : volume ratio = 6 : 1

Experiment B
2 cm
2 cm

surface area = (2 cm × 2 cm) × 6 faces = 24 cm²
volume = 2 cm × 2 cm × 2 cm = 8 cm³
surface area : volume ratio = 24 : 8 = 3 : 1

Experiment C
1 cm
1 cm

surface area = 6 cm² × 8 cubes = 48 cm²
volume = 1 cm³ × 8 cubes = 8 cm³
surface area : volume ratio = 48 : 8 = 6 : 1

In experiment A, a small agar cube is placed into water coloured with dye. The time it takes to completely change colour is short.

Experiment B uses a larger cube. It takes longer to completely change colour than the small cube. This is because even though its surface area is much larger, its surface area:volume ratio is smaller. It takes longer for the dye molecules to diffuse to the centre of the larger cube.

Experiment C uses eight small cubes. Their total volume is the same as the large cube, but their total surface area is much larger. This means the surface area:volume ratio is greater. The time taken for all eight cubes to completely change colour is as short as in experiment A.

How does this affect living organisms?

The surface area:volume ratio explains why larger organisms are made of so many small cells rather than one large one. It also explains why each individual cell is so small – if it was larger it would take too long for substances to move from the outside of the cell to its centre.

In addition, a larger organism needs exchange surfaces to provide it with a greater surface area:volume ratio. This enables it to exchange substances efficiently enough to support all of the cells in its body.

Questions

1 Write down two examples of exchange surfaces in multicellular organisms.

2 The lungs provide an exchange surface in the human body.
   a Explain how the structure of the lungs is adapted to their function as an exchange surface.
   b Explain why it is necessary for the human body to have exchange surfaces.

3 A cube measures 3 cm along every edge. What is its surface area:volume ratio?

4 Look at the amoeba in the photograph at the start of this section. It is an irregular shape with lots of finger-like protrusions. Suggest one advantage to the amoeba of having this shape.

B5.2 How does the nervous system help us respond to changes?

A: Detecting changes and responding

When you exercise, your body temperature rises and you feel hot. You start to sweat. A change in your environment, like an increase in temperature, is called a **stimulus**. Sweating is a **response** to the change in temperature. Eating is a response to the stimulus of hunger. Scratching is a response to the stimulus of an itch.

The way an animal responds to changes in its surroundings is important for its survival.

Detecting changes
Receptors
You can only respond to a change if you can detect it. **Receptors** inside and at the outside surface of your body detect stimuli.

You can detect many different stimuli, for example, sound, texture, smell, temperature, and light. Different types of receptors detect different types of stimulus. Receptors at the outside surface of your body monitor the external environment. Receptors inside your body monitor the internal environment of your body, for example, core temperature and blood sugar level.

Sense organs
Some receptors are made up of single cells, for example, pain receptors in your skin. Other receptor cells are grouped together as part of a complex sense organ, for example, your ear. Hearing is very important in humans and most other mammals. Detecting sounds gives you information about your surroundings.

The human ear is a sense organ. Sound waves entering the ear cause the ear drum to vibrate. The vibrations are transmitted along small bones to the cochlea. Receptor cells in the cochlea send nerve impulses to the brain.

Find out about
- the components of the nervous system, including receptors, neurons, and effectors
- the peripheral nervous system and the central nervous system
- the structure of neurons
- synapses

This woman's body is sweating in response to feeling too warm.

Key words
➤ stimulus
➤ response
➤ receptor

Responding to changes

The body's responses to stimuli are carried out by **effectors**. In multicellular organisms, muscles are effectors. Muscles can be made to contract or relax to help the body respond to a stimulus. For example, muscles in the hand and arm contract to pull your hand away from a hot object. Muscles in the eye contract and relax to change the shape of the lens, to focus light from near and far objects.

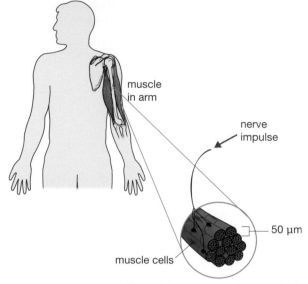

Muscles are one type of effector. They respond to stimuli by contracting or relaxing.

The nervous system

Receptors and effectors around your body are joined to your brain and spinal cord by nerves. Together, all of these parts form your nervous system. The brain and spinal cord are the **central nervous system (CNS)**. The CNS is linked to receptors and effectors elsewhere in the body by nerves – this is sometimes called the **peripheral nervous system (PNS)**.

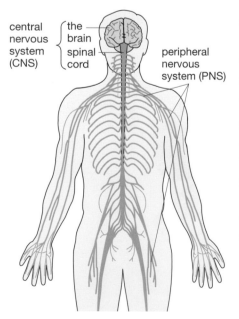

The development of a nervous system allowed larger and more complex multicellular animals to evolve. The nervous system enabled these animals to better respond to the environment. This gave them a survival advantage over simpler animals.

In humans and other mammals, receptors and effectors in the PNS are joined to the CNS (the brain and spinal cord) by neurons (nerve cells).

Nerves and neurons

Nerves are bundles of specialised cells called **neurons**. Like most body cells, neurons have a nucleus, a cell surface membrane, and cytoplasm. They are different from other cells because the cytoplasm is shaped into a very long, thin extension. This is called the **axon**, and it is how neurons connect different parts of the body.

Neurons carry **nerve impulses**. These allow the different parts of the nervous system to communicate with each other. The nerve impulse is carried along the axon, which in most neurons is insulated by a **fatty sheath** wrapped around the outside of the cell. The fatty sheath increases the speed that impulses move along the axon.

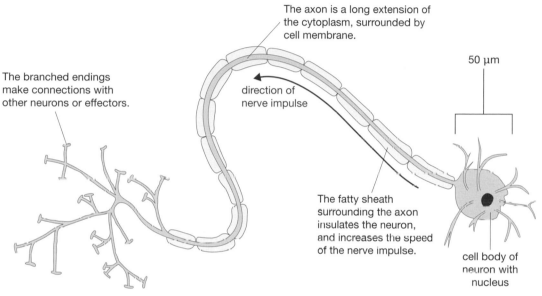

The axon is a long extension of the cytoplasm, surrounded by cell membrane.

50 µm

The branched endings make connections with other neurons or effectors.

direction of nerve impulse

The fatty sheath surrounding the axon insulates the neuron, and increases the speed of the nerve impulse.

cell body of neuron with nucleus

The structure of a motor neuron.

There are two different kinds of neurons:

- **Sensory neurons** carry nerve impulses from receptors to the CNS.
- **Motor neurons** carry nerve impulses from the CNS to effectors.

Your CNS coordinates all the information it receives from your receptors via sensory neurons. Information about a stimulus goes to either your brain or to your spinal cord. Motor neurons stimulate effectors to carry out the necessary response.

Key words

➤ synapse
➤ transmitter substance

When you play a computer game you need fast reactions to win. Nerve impulses give you fast reactions because they travel along axons at 400 metres per second.

Fast reactions are essential for survival. They help predators catch prey, and help prey escape!

Mind the gap

Neurons do not touch each other. So when nerve impulses pass from one neuron to the next, they have to cross tiny gaps. The gap is called a **synapse**. Some drugs and poisons (toxins) interfere with nerve impulses crossing a synapse – this is how they affect the human body.

How do nerve impulses cross a synapse?

Nerve impulses cannot jump across a synapse. Instead, molecules of **transmitter substances** are used to pass an impulse from one neuron to the next.

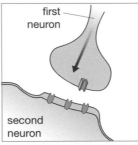 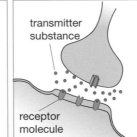 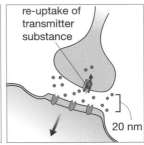

A nerve impulse arrives at a synapse. The direction of the impulse is shown by the arrow.

A transmitter substance is released from the first neuron. It diffuses across the synapse. The substance is the correct shape to fit into receptor molecules on the cell surface membrane of the second neuron.

A nerve impulse is stimulated in the second neuron. The transmitter substance is absorbed back into the first neuron to be used again.

How a synapse works.

Do synapses slow down nerve impulses?

The gap at a synapse is only about 20 nanometres wide. The synapse chemical travels across this gap in a very short time. Synapses do slow down nerve impulses to about 15 metres per second. But a nerve impulse still travels from one part of your body to another at an incredible speed.

Questions

1 Hamit steps outside into bright sunlight. He closes his eyes to protect them from the bright light. In this situation:
   a what is the stimulus?
   b what is the response?

2 When you type a text message to a friend, you are using receptors and effectors.
   a In which part of the body are these receptors?
   b In which part of the body are these effectors?

3 Describe the difference between:
   a the CNS and the PNS
   b an axon and a neuron
   c the jobs of a sensory neuron and a motor neuron.

4 Draw a flowchart to describe how a nerve impulse is passed from one neuron to another across a synapse.

B: Reflex responses

Your nervous system enables you to respond very quickly to changes in your environment. Many responses happen automatically without you having to think about them. For example, when a bright light shines in your eye, your pupil becomes smaller. This stops bright light from damaging the sensitive receptor cells at the back of your eye. This kind of response is a **reflex response**. Many reflex responses are crucial to survival.

Human reflexes

In a simple reflex, nerve impulses are passed along a pathway called a **reflex arc**. The diagram below shows this pathway for a pain reflex. A **relay neuron** in the spinal cord connects the sensory neuron to the motor neuron without linking to the brain.

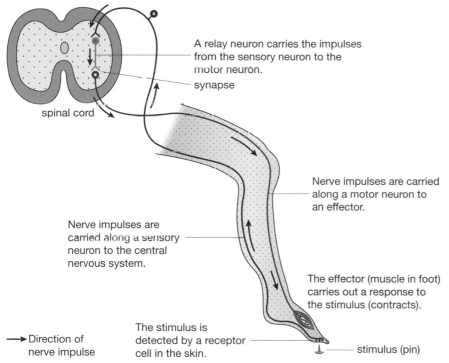

The CNS is made up of the brain and spinal cord. It coordinates the body's responses.

A relay neuron carries the impulses from the sensory neuron to the motor neuron.

synapse

spinal cord

Nerve impulses are carried along a motor neuron to an effector.

Nerve impulses are carried along a sensory neuron to the central nervous system.

The effector (muscle in foot) carries out a response to the stimulus (contracts).

→ Direction of nerve impulse

The stimulus is detected by a receptor cell in the skin.

stimulus (pin)

A summary of a reflex arc. (Note: not drawn to scale.)

Many reflex arcs do not include the brain. This means reflexes are involuntary – they happen automatically and you do not have to think about them.

Conscious control of reflexes

Most human reflex arcs are coordinated by the spinal cord. A reflex arc only has a simple connection between a sensory neuron and a motor neuron. Your brain does not have to make a decision. The response happens automatically, because the reflex is adapted to help you survive.

But sometimes a reflex may not be what you want to happen. Some reflexes can be modified by conscious control. Imagine picking up a hot plate.

Find out about

- reflex responses
- the structure of a reflex arc

Key words

➤ reflex response
➤ reflex arc
➤ relay neuron

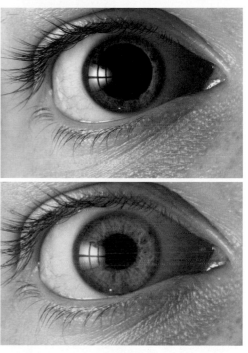

When the light is switched on in a dark room, muscles in the iris of your eye cause your pupil to shrink. This reflex response protects the eye from damage caused by the sudden bright light.

When you put your finger in a newborn baby's open palm, the baby grips the finger. When you pull away, the grip gets stronger. This is the grasping reflex. This reflex response usually disappears by age six months.

Your pain reflex would make you drop it. But if your dinner is on the plate, you can override this reflex and hold onto the plate until you put it down safely. The conscious control of your brain overrides the reflex response. The diagram below explains how this happens.

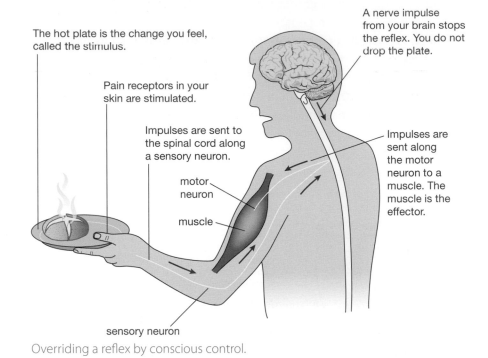

The hot plate is the change you feel, called the stimulus.

Pain receptors in your skin are stimulated.

Impulses are sent to the spinal cord along a sensory neuron.

motor neuron

muscle

sensory neuron

A nerve impulse from your brain stops the reflex. You do not drop the plate.

Impulses are sent along the motor neuron to a muscle. The muscle is the effector.

Overriding a reflex by conscious control.

Reflexes in simple animals

Simple animals always respond to a stimulus in the same way. For example, woodlice always move away from light. This is an example of a simple reflex response. Simple reflex behaviour increases the animal's chance of survival. It helps the animal to:

- find food, shelter, or a mate
- escape from predators
- avoid harmful environments, for example, extreme temperatures.

Simple reflexes usually help animals to survive. But animals that only behave with simple reflexes cannot easily change their behaviour or learn from experience. This could be a problem if conditions around them change. Their simple reflexes may no longer be helpful for survival.

Woodlice move away from light, so you are most likely to find them in dark places. Their behaviour is all based on reflexes.

Questions

1 Write down two examples of reflexes in humans.

2 Which of the following is most likely to be a reflex response?
   A lying down when you're tired
   B blinking when something comes near your eye
   C putting on a jumper when it's cold
   D eating when you feel hungry.

3 Draw a labelled diagram to show a reflex arc for the grasping reflex in a newborn baby.

A: The endocrine system

As you know, the body's responses to stimuli are carried out by effectors. Your muscles are one type of effector. But your body has another type of effector – **glands**. These are organs that make substances called **hormones** and release them into the blood. Hormones are transported by the blood and affect other cells and tissues. The presence – and absence – of hormones causes changes in the body.

Find out about

- how hormones control responses in the body
- H the roles of adrenaline and thyroxine
- negative feedback

Key words

➤ gland
➤ hormone
➤ secretion
➤ ADH

Julie is in her fifties. Her body has stopped producing certain hormones. As a result, her body is changing. This is called the menopause. Symptoms of the menopause include hot flushes.

Glands

A change in the body's external or internal environment is detected as a stimulus by a receptor. A nerve impulse is sent to the CNS along a sensory neuron. A nerve impulse is then sent along a motor neuron to a gland. This causes the gland to release hormone molecules into the blood (a process called **secretion**).

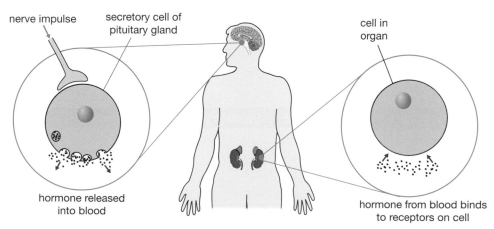

nerve impulse

secretory cell of pituitary gland

cell in organ

hormone released into blood

hormone from blood binds to receptors on cell

The diagram illustrates how the pituitary gland in the brain controls cells in the kidneys using a hormone called **ADH**.

Together, glands and hormones make up the **endocrine system**. There are glands in various places in the body.

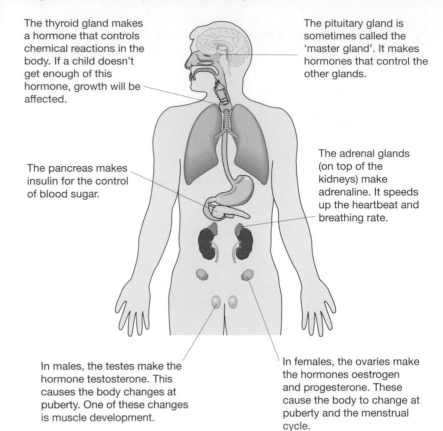

The thyroid gland makes a hormone that controls chemical reactions in the body. If a child doesn't get enough of this hormone, growth will be affected.

The pituitary gland is sometimes called the 'master gland'. It makes hormones that control the other glands.

The pancreas makes insulin for the control of blood sugar.

The adrenal glands (on top of the kidneys) make adrenaline. It speeds up the heartbeat and breathing rate.

In males, the testes make the hormone testosterone. This causes the body changes at puberty. One of these changes is muscle development.

In females, the ovaries make the hormones oestrogen and progesterone. These cause the body to change at puberty and the menstrual cycle.

Some important glands in the human body.

Hormones

Hormones are sometimes called chemical messengers. They allow different parts of the body to communicate without using nerves. Hormones are transported around the body in the blood. Different organs and tissues around the body have receptors for specific hormones. When the correct hormone binds to the receptors, the organ or tissue responds by behaving in a particular way.

Hormones can bring about changes in cells or tissues that are a long way away from the gland that secreted the hormone. Hormones move through the circulatory system less quickly than nerve impulses move through the nervous system, but the effects brought about by hormones tend to last longer.

Response caused by:	Speed of response	Duration of response	Example
nerve impulses	fast	short-lived	muscle contraction (e.g., to pull your hand away from a hot object)
hormones	not as fast	longer-lasting	blood sugar level gradually lowered

Key words

➤ endocrine system
➤ thyroxine
➤ thyroid gland
➤ adrenaline
➤ adrenal gland
➤ pituitary gland
H ➤ negative feedback

One important hormone in the human body is **thyroxine**, secreted by the **thyroid gland**. It regulates growth and the rates of chemical reactions in cells.

Another important hormone is **adrenaline**. This hormone is secreted by the **adrenal gland** in large quantities at times of stress. It prepares the body for action by:

● increasing the heart rate and breathing rate

● causing the liver to break down stored carbohydrate to release glucose (for cellular respiration)

● stimulating muscle contraction.

A surge of adrenaline in response to a threat causes what is sometimes called the 'fight-or-flight' response.

Glands working together

The **pituitary gland** in the brain is sometimes called the 'master gland'. It makes hormones that control other glands. For example, it makes a hormone called thyroid stimulating hormone (TSH), which causes the thyroid gland to make thyroxine.

How is hormone production switched off?

Hormone production is switched on in response to a stimulus. However, if a gland continued to make a hormone for a long time it could be very dangerous. The effects triggered by the hormone could eventually damage the body. So hormone production must be switched off.

Often, a hormone produced by a gland not only affects the target organ, but also causes the gland to stop making the hormone. This is the case for thyroxine. It is an example of **negative feedback**.

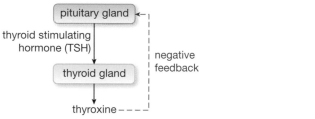

When stimulated, the thyroid gland makes thyroxine. Thyroxine causes the pituitary gland to stop making TSH. Without TSH, the thyroid gland is not stimulated to make any more thyroxine.

Questions

1 What is the name of the system that uses glands and hormones?

2 Explain how a gland in the brain could control the behaviour of cells in another organ.

3 Which gland secretes each of the following hormones?

a ADH b insulin c adrenaline d thyroxine.

H 4 A baby is born with a faulty pituitary gland. It does not make TSH. Suggest how this is likely to affect the baby.

A: Changing to stay the same

Find out about

- the importance of maintaining a constant internal environment
- homeostasis

When you feel ill, your parents or a doctor might take your body temperature. They're looking to see whether it's getting too high. But why is this so important? Your core body temperature is the temperature of your vital internal organs. If it gets too high or low you can become very unwell and even die.

It's not just your body temperature that's important. The water balance of your body (how hydrated you are) and your blood sugar level also affect how well your body functions. Your body constantly works to keep its internal environment the same. This is called **homeostasis**.

Exercise can make you feel very hot. Your body sweats more to help cool you down. The sweating response is an example of homeostasis.

Optimum conditions

You learnt in B3.1B that enzymes work best within a very narrow **range** of conditions. For example, the rate of an enzyme-catalysed reaction will be fastest at a particular temperature. This is the enzyme's optimum temperature. Cells, enzymes, and life processes work best in particular conditions.

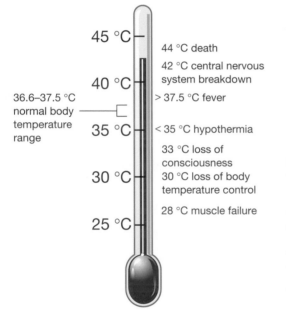

45 °C

44 °C death

42 °C central nervous system breakdown

40 °C

> 37.5 °C fever

36.6–37.5 °C normal body temperature range

35 °C

< 35 °C hypothermia

33 °C loss of consciousness

30 °C

30 °C loss of body temperature control

28 °C muscle failure

25 °C

Your core body temperature is about 37 °C, but a small variation either side of this is normal. A core temperature over 42 °C or under 28 °C usually results in death.

Homeostasis is the body working to counteract changes. If you get too hot, your body works to cool you down. If you drink too much water, your body works to get rid of some of it. Homeostasis relies upon:

- receptors – to detect a change (a stimulus)
- the CNS – to receive information and coordinate a response
- effectors – to action the response to help reverse the change.

Receptors send messages to the CNS along sensory neurons. The CNS sends messages to effectors, including muscles and glands, along motor neurons. Glands send messages to other effectors by secreting hormones.

A change in the internal conditions of the body results in actions that reverse the change back to a steady state. This is negative feedback. Negative feedback systems are all around you. For example, if the temperature inside your fridge goes up, the motor switches on to cool it back down to the original temperature. When it is cool enough, the motor switches off.

Keeping conditions constant

Keeping any system in a constant, steady state can be tricky. You can make a simple model of homeostasis in the laboratory.

A model of homeostasis

A beaker of water can be used as a very simple model of the human body (approximately 60% of an average adult human body is water).

Use a clamp to hold a thermometer so that the bottom of the thermometer is in the middle of the water. Use a Bunsen burner to heat the water to exactly 37 °C. Now try to keep the water at exactly 37 °C by removing and replacing the Bunsen burner as necessary. Is it easy to keep the water at a constant temperature?

Now do the experiment again. This time you can use the Bunsen burner and a supply of ice cubes to try to keep the water temperature at 37 °C. Is it easier this time?

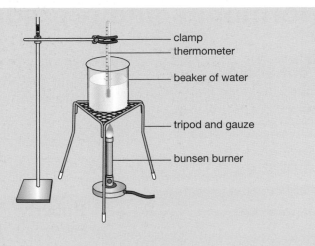

clamp
thermometer
beaker of water
tripod and gauze
bunsen burner

In the model, the Bunsen burner and the ice cubes are **antagonistic** effectors. They have opposite effects. One increases the temperature of the water, and the other decreases it. Your body uses antagonistic effectors to keep its internal environment constant.

In the model, you had to think about which effector to use and when. But in your body homeostasis happens automatically.

Key words

➤ homeostasis
➤ range
➤ antagonistic

Questions

1 Your body constantly works to keep its internal environment the same. What is this process called?

2 Choose the correct word to complete each sentence.
   a When a change causes actions that reverse the change, this is called *effective/negative/positive* feedback.
   b A change in the body's environment is detected by *detectors/effectors/receptors*.
   c Messages are sent to the CNS through *glands/motor neurons/sensory neurons*.
   d Effectors that have opposite effects are described as *antagonistic/constant/optimum*.

3 The experiment in the model box is a simple model of homeostasis.
   a In the model, which part represents:
      i a receptor?
      ii the CNS?
      iii an effector?
   b Suggest why an electronic water bath is a better model of homeostasis.

A: Hormones and reproduction

Find out about

- the role of hormones in human reproduction
- **H** how hormones interact to control the menstrual cycle
- use of hormones in contraception
- **H** use of hormones to treat infertility

Key words

➤ menstrual cycle
➤ oestrogen
➤ testosterone

Humans – like all organisms – need to reproduce to ensure the survival of the species. Puberty plays an important role in preparing human females and males for sexual reproduction. During this time, hormones cause lasting physical changes in the body. Hormones continue to be important in adults, including controlling the **menstrual cycle** in females. Without these processes, sexual reproduction would not be possible.

Puberty

Puberty is the process that changes a child's body into an adult body capable of sexual reproduction. The process usually starts around age 10 and is usually complete by age 17, but this varies from person to person. Two hormones are very important during puberty: **oestrogen** and **testosterone**.

- In females, oestrogen causes thickening of the lining of the vagina and uterus (womb), widening of the hips, and growth of breasts.

- In males, testosterone causes growth of tissues in the testicles that make sperm and growth of the penis.

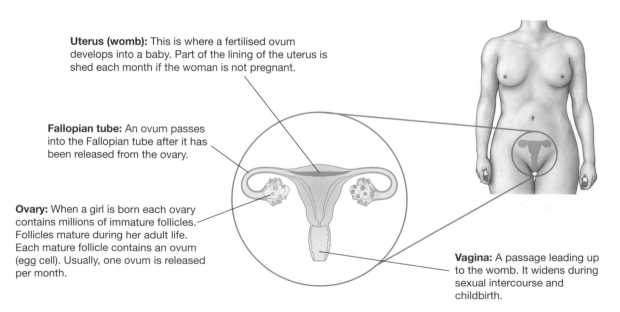

Uterus (womb): This is where a fertilised ovum develops into a baby. Part of the lining of the uterus is shed each month if the woman is not pregnant.

Fallopian tube: An ovum passes into the Fallopian tube after it has been released from the ovary.

Ovary: When a girl is born each ovary contains millions of immature follicles. Follicles mature during her adult life. Each mature follicle contains an ovum (egg cell). Usually, one ovum is released per month.

Vagina: A passage leading up to the womb. It widens during sexual intercourse and childbirth.

The female reproductive system.

The menstrual cycle

The menstrual cycle is controlled by hormones. It includes the important process of **ovulation**, in which an ovum (egg cell) is released from the ovaries. This is essential for sexual reproduction. Following ovulation, the lining of the uterus thickens, ready to receive a fertilised ovum. If the ovum is not fertilised the thickened lining of the uterus breaks down and is shed. This is **menstruation**.

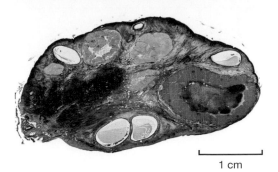

A section through a human ovary, seen under a light microscope. The white ovals are follicles. The pink structure is a mature follicle that is breaking down after ovulation.

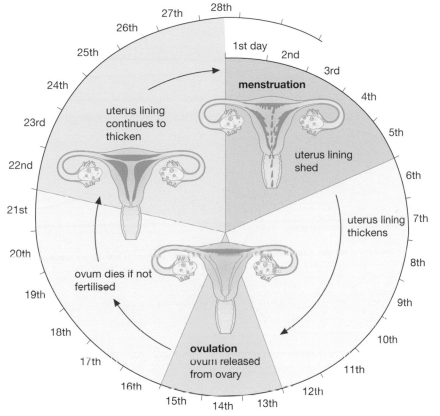

The menstrual cycle. Menstruation starts at puberty and stops when the woman is between about 45 and 55 years old. This time in her life is the menopause. After this time, a woman can no longer have children.

Interacting hormones

A number of hormones interact to control the menstrual cycle, including:

- **Follicle-stimulating hormone (FSH)** and **luteinising hormone (LH)** – secreted by the pituitary gland
- oestrogen and **progesterone** – secreted by the ovaries.

The effects of these hormones during the menstrual cycle are summarised on the next page.

At the end of the 28-day cycle progesterone levels fall and the thickened uterus wall begins to break down. It is discharged in blood through the vagina (menstruation) as the cycle starts again.

Key words

- ➤ ovulation
- ➤ menstruation
- ➤ follicle-stimulating hormone (FSH)
- ➤ luteinising hormone (LH)
- ➤ progesterone

Luton Sixth Form College
Learning Resources Centre

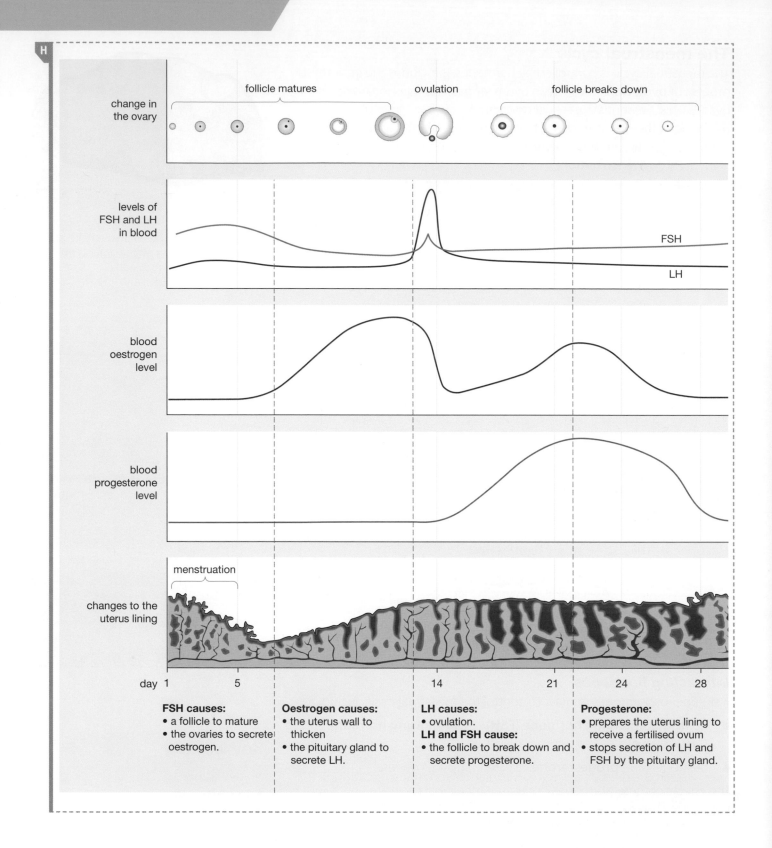

change in
the ovary

follicle matures

ovulation

follicle breaks down

levels of
FSH and LH
in blood

FSH

LH

blood
oestrogen
level

blood
progesterone
level

changes to the
uterus lining

menstruation

day 1   5   14   21   24   28

FSH causes:
• a follicle to mature
• the ovaries to secrete
  oestrogen.

Oestrogen causes:
• the uterus wall to
  thicken
• the pituitary gland to
  secrete LH.

LH causes:
• ovulation.
LH and FSH cause:
• the follicle to break down and
  secrete progesterone.

Progesterone:
• prepares the uterus lining to
  receive a fertilised ovum
• stops secretion of LH and
  FSH by the pituitary gland.

Controlling reproduction

Contraception

Hormones can be used as a form of contraception to stop a woman getting pregnant. Often these hormones are swallowed in a **contraceptive pill**. Different types of contraceptive pill contain different combinations of hormones, which prevent pregnancy by disrupting the menstrual cycle. For example, one type of contraceptive pill contains progesterone, which stops the secretion of LH. Without LH, ovulation does not happen and no ovum is released.

Contraceptive pills reduce the risk of pregnancy but they do not reduce the risk of being infected with (or passing on) sexually transmitted infections (STIs). Infections such as HIV, chlamydia, gonorrhoea, genital warts, and others can be passed between people who are only using a contraceptive pill. Only a physical barrier such as a condom can reduce the risk of pregnancy and STIs.

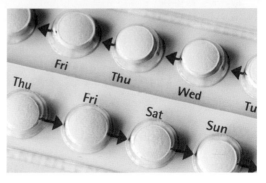

Contraceptive pills contain hormones. They are used to disrupt the menstrual cycle, for example to prevent ovulation, which stops a woman from getting pregnant.

H Treating infertility

Sadly, some women find it very difficult to get pregnant. They may be told by a doctor that they are infertile. However, hormones can be used to treat certain cases of female **infertility**. In these cases, the cause of infertility is that a hormone that regulates the menstrual cycle is not working properly. This could mean that woman does not ovulate. Giving the correct hormone at the correct time can help ovulation, and other stages of the menstrual cycle, to happen correctly.

Hormones are also used in a fertility treatment called *in vitro* fertilisation (IVF). Hormone injections are used to trigger ovulation. Often several eggs are released, which are collected. The eggs are fertilised using sperm in a glass container. More hormone treatment prepares the uterus lining to receive a fertilised egg. Fertilised eggs are then implanted into the uterus through the vagina.

The use of hormones to treat infertility is an example of an application of science that has made a significant positive difference to people's lives.

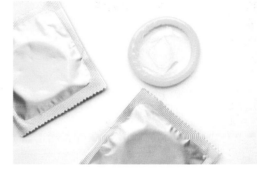

Unlike contraceptive pills, condoms are physical barriers that prevent the spread of sexually transmitted diseases.

Key words

> contraceptive pill

H > infertility

Questions

1 Describe two ways in which hormones are important in preparing human beings to reproduce sexually.

2 What is the difference between ovulation and menstruation?

H 3 Explain how the hormones FSH, LH, oestrogen, and progesterone interact to control the menstrual cycle.

4 Thora heard that a contraceptive pill can prevent pregnancy. What else should she consider when choosing a method of contraception?

A: Controlling blood sugar levels

Find out about

- how hormones control blood sugar levels
- what causes diabetes
- how diabetes can be treated

Key words

- ➤ insulin
- H ➤ glucagon
- ➤ type 1 diabetes
- ➤ type 2 diabetes

Fizzy drinks and sweets can contain high levels of sugar, which is absorbed into the blood quickly. Rice and pasta contain complex carbohydrates that are broken down into sugar slowly, so you absorb it gradually.

Your body depends on sugar for cellular respiration, and without it you would die. Sugar is absorbed into the blood from the food you eat. Controlling the amount of sugar in the blood is very important. High blood sugar can make you feel very thirsty, tired, and produce large volumes of urine containing sugar. Low blood sugar can make you tremble; feel dizzy, tired, and irritable; and have difficulty concentrating. If it drops low enough, you can become drowsy or even lose consciousness.

Blood sugar level

Many processed foods have extra sugar added to them to improve their taste. The sugar is absorbed quickly into the blood. Consuming a lot of sugar in one go causes a surge of the hormone **insulin** to remove sugar from the blood. This may cause your body to remove too much sugar from the blood, so the sugar high is followed by a sugar low.

Foods that are high in fibre and complex carbohydrates are better for you. The carbohydrates in foods such as pasta, rice, and bread are digested slowly. Their sugars are released and absorbed gradually into your blood, so it's easier for your body to keep your blood sugar level balanced.

When blood sugar level is high, insulin is secreted by the pancreas and causes the absorption of blood sugar by cells. When blood sugar H level becomes too low, the hormone **glucagon** is secreted by the pancreas and causes the liver to break down stored carbohydrate to release glucose.

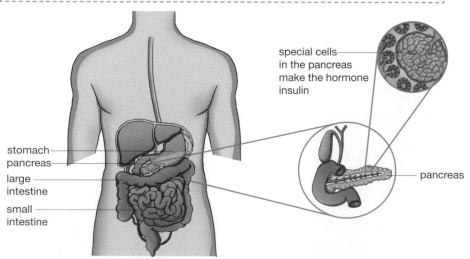

Blood sugar level is controlled using hormones secreted by cells in the pancreas.

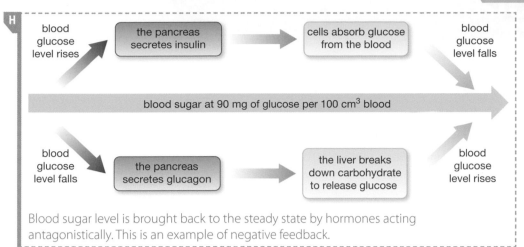

blood sugar at 90 mg of glucose per 100 cm³ blood

Blood sugar level is brought back to the steady state by hormones acting antagonistically. This is an example of negative feedback.

Type 1 diabetes can be treated with daily insulin injections and careful diet.

Diabetes

People with diabetes cannot control their blood sugar level. There are two main types of diabetes: type 1 and type 2.

Type 1 diabetes

In people with **type 1 diabetes** the pancreas stops making enough insulin. The blood sugar level can rise to dangerously high levels, particularly after a sugary meal.

People with type 1 diabetes need several daily injections of insulin to control their blood sugar level. They have to be careful about what they eat, to match their sugar intake to their lifestyle. They may have to prick their finger each day and measure their blood sugar level using a test strip.

As you found out in B4.4, stem cell treatment could help to replace insulin-secreting cells in the pancreas. This would help people with type 1 diabetes to make their own insulin without needing injections. This kind of treatment is still being tested in clinical trials, and we do not yet know whether or not its potential benefits will be outweighed by its risks.

Type 2 diabetes

People with a poor diet, inactive lifestyle, or who are obese may develop **type 2 diabetes** (sometimes called 'late-onset diabetes'). The number of young people with type 2 diabetes is rising. In type 2 diabetes the body gradually stops making enough insulin for your needs or your cells stop being able to use the insulin properly.

People with type 2 diabetes need to take regular, moderate exercise to control their blood sugar levels. They have to eat carefully and plan their diet, so that sugar is absorbed into the blood gradually.

Long-term effects

There is presently no cure for either type of diabetes. Both types can cause lasting damage to the body. For example, as you found out in B2.4D, diabetes is a risk factor for cardiovascular disease. Over time 'hardening of the arteries' takes place, which can lead to heart attacks, kidney damage, or sight problems (where blood vessels in the retina are involved).

Some diabetics wear a wrist band indicating that they have diabetes. Why do you think they do this?

Questions

1 Make a table to compare the causes, symptoms, and treatments for type 1 and type 2 diabetes.

2 Explain why exercise causes the blood sugar level to fall.

3 Your classmate feels faint after PE. You notice that she is wearing a wrist band showing that she is diabetic. You rush to tell your teacher. What do you think might have happened? What would you suggest that the teacher does to help?

4 Suggest why insulin has to be injected. Why can't it be swallowed like other medicines?

Science explanations

B5 The human body – staying alive

In this chapter you learnt about how systems in the human body work together to help us stay alive by supplying cells, removing waste, and detecting and responding to changes. Understanding the nervous and endocrine systems has helped us to control reproduction, treat infertility, and offer hope to those with damaged sense organs and neurons.

You should know:

- about substances transported into, out of, and around the human body, including oxygen, carbon dioxide, water, dissolved food molecules, and urea
- how the human circulatory system interacts with the gaseous exchange system, the digestive system, and the excretory system
- how the structures of the heart, arteries, veins, capillaries, red blood cells, and blood plasma are adapted to their functions
- about surface area volume ratio, and why exchange surfaces are necessary in multicellular organisms
- how the components of the nervous system work together to enable it to function
- how the structures of nerve cells, synapses, and reflex arcs relate to their functions
- the principles of hormonal coordination and control by the human endocrine system
- why it's important for the body to maintain a constant internal environment
- **H** about the principle of negative feedback
- about the roles of hormones in human reproduction, including the control of the menstrual cycle
- how hormones can be used as contraceptives, and their risks and benefits
- **H** about the use of hormones to treat infertility
- how blood sugar levels are controlled in the body
- the differences between type 1 and type 2 diabetes, and how they are treated.

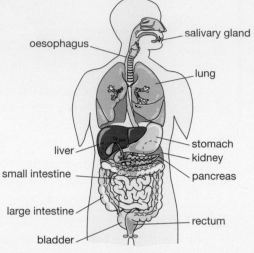

Organs in the body work together in systems, which cooperate to support the life processes of our cells and of the body as a whole.

Ideas about Science

Applications of science can help us to improve our quality of life. For example, using hormones to control the menstrual cycle can help us to prevent pregnancy by preventing ovulation. Contraception that only uses hormones reduces the risk of pregnancy but does not reduce the risk of passing on sexually transmitted infections (STIs).

We can also use hormones to treat some types of infertility. Our understanding of stem cells offers the hope of treatments for diabetes. These applications of science can make a positive difference to people's lives. However, nothing is risk free, and new technologies can introduce new risks. Some applications of scientific knowledge have ethical implications, for example, when stem cells are sourced from embryos.

You should be able to:

- identify risks associated with a course of action, or that have arisen from a new scientific or technological advance

- interpret and discuss information on the size of a given risk, taking account of both the chance of it occurring, and the consequences if it did

- suggest reasons for people's willingness to accept a risk

H • distinguish between perceived and calculated risk

- identify ethical issues and summarise different views that may be held.

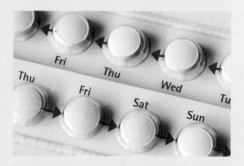

Contraceptive pills contain hormones that prevent ovulation and stop a woman getting pregnant. But they do not prevent the spread of STIs.

B5 Review questions

1 **a** How do each of the following substances move through partially permeable membranes in the human body?

     i carbon dioxide   **ii** urea   **iii** glucose.

  b What is the name of the substance that moves through partially permeable membranes by osmosis?

2 When you breathe in, molecules of oxygen from the air end up in your blood. Describe the path to the blood taken by a molecule of oxygen from the air entering the nose. Your answer should include which structures it passes through and how it moves across an exchange surface.

3 Copy and complete the table for the four main components of blood.

Blood component	Function	How the component is adapted to its function
red blood cells		
		ability to make antibodies and ingest pathogens
platelets		
		acts as a solvent

4 Simple single-celled pond organisms exchange oxygen, food, and waste with the surrounding water. Explain why you need a circulation system, whereas simple pond creatures do not.

5 Mary is learning about communication systems in the human body. She reads about the nervous system and the endocrine system. She looks at a model that describes nerve impulses as being like sending an email, and hormones like sending a letter.

  Suggest what aspects of the two systems are represented by Mary's model.

6 Explain why it is important to maintain a constant internal environment in the human body.

7 The diagram shows a very simple model of the menstrual cycle.

  Four stages have been labelled **A**, **B**, **C** and **D**.

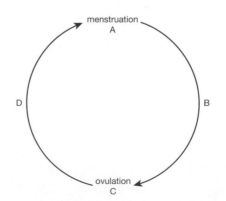

a The following statements describe what happens during these four stages in the menstrual cycle.
The statements are *not* in the correct order.
Write the correct letter, **A**, **B**, **C** or **D**, next to each stage in the table.

Description	Letter
uterus lining thickens	
follicle breaks down, and uterus lining ready to receive fertilised ovum	
uterus lining is shed	
ovum released from follicle in ovary	

H **b** Which hormone causes the uterus wall to begin to thicken?

c Which hormone causes ovulation?

8 The human body works to control its blood sugar level.

H **a** Choose words from the list to complete the flow diagrams to explain how blood sugar level is controlled.
Each word may be used once, more than once, or not at all.

carbohydrate   fat   glucagon

glucose   insulin   protein

| blood glucose level *higher* than normal |
| ↓ |
| pancreas secretes .... |
| ↓ |
| cells absorb .... from the blood |
| ↓ |
| blood glucose level returns to normal |

| blood glucose level *lower* than normal |
| ↓ |
| pancreas secretes .... |
| ↓ |
| liver breaks down .... to release .... |
| ↓ |
| blood glucose level returns to normal |

b In people with diabetes the body cannot control its blood sugar level.

Marjorie describes her condition.

 i Which type of diabetes does Marjorie have? Explain your answer.

 ii How could Marjorie's diabetes be treated?

Marjorie
It was shortly after my 60th birthday when the doctor told me I had diabetes. I've always been a bit overweight, but otherwise I've been fine. I didn't expect this!

B6 Life on Earth – past, present, and future

Why study evolution and biodiversity?

The great diversity of life on Earth has evolved over millions of years, and will continue to evolve for millions of years to come. Biodiversity is a precious resource that we depend on for our survival. But the survival of the human race is far from certain. We have damaged the Earth's biodiversity, and thousands of species have become extinct because of human actions. We may suffer the same fate unless we learn to take better care of life on Earth now to protect it for the future.

What you already know

- There are many different types of organisms on Earth, living in many different environments.

- Organisms are well adapted to the habitats they live in.

- Changes in the environment may leave organisms less well adapted, and this can lead to extinction.

- Living organisms have changed over time.

- Fossils provide information about organisms that lived between tens of thousands and millions of years ago.

- Living organisms produce offspring of the same kind, but normally offspring vary and are not identical to their parents.

- There is variation between individuals within a species.

- Variation can be described as continuous or discontinuous.

- Variation means some organisms compete more successfully, resulting in natural selection.

- Variation, adaptation, competition, and natural selection result in the evolution of species.

- Organisms can be grouped and classified based on their similarities and differences.

- Classification keys can be used to identify which species an organism belongs to.

- Some of the reasons why it's important to protect and conserve biodiversity, and some ways of doing this.

The Science

Our understanding of genomes helps us to explain how some helpful adaptations appear and can be passed on to offspring. We can make links between DNA, inheritance, variation, and natural selection to explain changes in species over many generations, and how new species are formed. Scientists collect evidence for evolution from fossils and DNA, and observe modern examples of evolution in the world around us. DNA evidence also helps us to classify species.

All living organisms depend on the Earth's biodiversity for their survival, but unfortunately it is being damaged by human activity. Thankfully, there are things we can all do to protect it for future generations.

Ideas about Science

Today most scientists accept the explanation of evolution by natural selection. But the explanation will change as we collect new evidence and test the ideas.

The Earth's biodiversity is a valuable resource, but it is threatened by human activity. Scientists are helping us devise ways to protect ecosystems and species, and to use natural resources in a more sustainable way.

B6.1 How was the theory of evolution developed?

A: The modern theory of evolution

Find out about
- the link between genetic variants and evolution
- competition, advantage, and natural selection
- the formation of new species

Key words
- ➤ species
- ➤ evolution
- ➤ natural selection
- ➤ variation
- ➤ advantage
- ➤ adaptation
- ➤ extinction
- ➤ mutation

It has been estimated that there are approximately 8.7 million **species** presently alive on Earth. Many millions of species that were alive in the past are now extinct. One thing that sets us apart as humans is our ability to ask questions about the world. We wonder, and try to explain, why the world is the way it is. Humans have been developing explanations about the origins of species for centuries.

Most scientists now are convinced that the theory of **evolution** by **natural selection** is the best explanation of how and why:

- species are well adapted to their environment
- new species are formed
- there are changes in species over time.

The modern theory of evolution is based on evidence from the natural world. It combines ideas about genomes, **variation**, **advantage**, competition, and inheritance to explain how the inherited characteristics of a population change over a number of generations.

Adapted for survival

All of the species presently alive on Earth are successful survivors. They have features, called **adaptations**, that help them survive in their environments.

Fish have adaptations that enable them to live in water. A fish absorbs oxygen dissolved in water. The oxygen diffuses from the water into the fish's blood across the large surface area of the gills. A streamlined body and a smooth surface help the fish move through the water with little resistance. Sets of fins keep the fish balanced. A swim bladder gives the fish buoyancy.

We can use the theory of evolution to explain how species get these adaptations.

Fish and cacti are both adapted to the environment in which they live. Fish are adapted to live under water. Cacti are adapted in different ways to live in the desert.

The importance of competition

All individual organisms compete with other organisms in their environment. They compete for resources including space, food, water, shelter, mates, light, pollinators, and seed dispersers. These resources are limited, so individuals have to compete for them.

There are differences between members of the same species. This variation means that an individual can have a feature that enables it to compete more successfully than other individuals – it has given them an advantage. Other features may also give individuals an advantage – for example, the ability to avoid or survive attack by predators. An individual that is better adapted to its environment than others in the population is said to have a better fit – hence the phrase 'survival of the fittest'.

Sometimes there are lasting changes in the conditions in an ecosystem. This affects the populations of organisms that live there. Within a population, some individuals may be better or less able to survive in the new conditions. If the conditions change too much, no organisms in the population may be able to survive. This can lead to **extinction**.

What causes variation?

You learnt in B1.1B that differences between individuals of the same species can be caused by genetic variants in their DNA. There is usually extensive genetic variation within a population. Differences can also be caused by the environment.

A genetic variant is created by a change in the sequence of bases in the DNA, called a **mutation**. Mutations can be caused by some substances and by certain types of radiation. However, most mutations are caused by errors when DNA is copied. Mutations, and the features they cause, occur at random. It is only by chance that mutations sometimes occur that cause useful features, which are then maintained by natural selection. Evolution does not plan in advance!

Most mutations have no effect on the organism – they do not change its phenotype. Mutations that do have an effect are usually harmful. Very rarely, a mutation causes a change that gives the organism an advantage. The mutation creates a genetic variant that gives rise to a phenotype better suited to the organism's environment. These mutations are rare, but they are important because they enable evolution to occur.

Passing it on

The changes in an organism within its own lifetime are not evolution. Evolution refers to changes in the inherited characteristics of a population over a number of generations.

A helpful feature can only be passed on to offspring if there has been a change in the organism's DNA. Additionally, only changes in the DNA of gametes are passed on.

Individuals with an advantage are more likely to survive longer. This means they are more likely to reproduce, and pass helpful genetic variants to their offspring. This is called natural selection.

All of these gulls are competing for the same food (worms) in the soil.

A gull with a slightly longer beak may be better at catching worms than other gulls in the population.

A mutation in a gene controlling fur colour can produce tigers with white fur.

① Living things in a species are not identical. They have variation.

Ancestors of modern giraffes had variation in the length of their necks.

② They compete for things like food, shelter, and a mate. But what if something in the environment changes?

Food supply became scarce. The giraffes competed for food.

③ Some will have features that help them to survive. They are more likely to breed. They pass their genes on to their offspring.

Taller giraffes were able to eat more food, so were more likely to survive and breed. They passed on their features to the next generation.

④ More of the next generation have the useful feature. If the environment stays the same, even more of the following generation will have the useful feature.

Over many generations, more giraffes with longer necks were born. The proportion of taller giraffes in populations increased.

The process of natural selection.

Because of natural selection, helpful genetic variants (and the features they cause) become more common in the population with each generation. Combinations of helpful genetic variants (and features) accumulate in offspring over many generations. Eventually the organisms within the population can be quite different from their ancestors. This explains how species change and become well adapted to their environment over time.

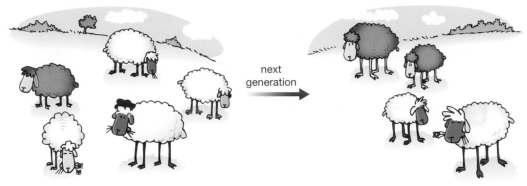

next generation

There is variation within every species. Natural selection means that some features become more common in a population in later generations.

All species, including humans, have evolved over many generations, and will continue to evolve as long as the species exists. We must not assume that evolution has finished, or that any species – including humans – is in its final (or perfect) form.

New species

The organisms in a population may evolve to be quite different from their ancestors, and from other populations of the same species. In this case, we may classify them as a new species.

Scientists define a species as a group of organisms so similar that:

● they can breed together

● their offspring can also breed (they are fertile).

horse

donkey

mule

Horses and donkeys look very similar. But their offspring (mules) are infertile. So horses and donkeys are different species.

A new species is more likely to form when a population of organisms is isolated. This means the isolated population has no contact with other groups of the same organism.

Mutations will occur at random in the individual genomes in the population. The individuals in the population can only breed with one another, because the population is isolated. So the new genetic variants will not be passed outside the population. Environmental conditions may be (or become) different in the ecosystem where the isolated population lives. Due to natural selection, the population will become adapted to its own unique environment. Eventually, the isolated population may be so different to other groups that we would classify it as a new species.

Questions

1 Explain why horses and donkeys are different species.

2 Look at the photograph of the cactus. Explain one way it is adapted to survive in its environment.

3 Look at the photograph of the two tigers. Their furs are different colours.
   a What caused the different fur colours?
   b The white fur is an example of variation. Could this variation be passed to the tiger's offspring? Explain your answer.
   c Which of the two tigers would have an advantage if their environment changed to become much colder and snowy?
   d If the environmental change was permanent, how would you expect the species to change? Explain why you think this.

4 Research suggests that early humans fought each other for resources. Fossils of early human skeletons commonly have broken bones in their faces. Later humans have much stronger bones in their faces. Suggest whether each of the following statements is TRUE or FALSE:
   a The changes in human face bones are an example of evolution.
   b The face bones mutated to become stronger.
   c Humans with stronger face bones had an advantage.
   d Humans needed to survive fights, so they evolved stronger face bones.

B: A theory developed from evidence

Find out about

- the impacts of selective breeding on the theory of evolution, and on our lives
- how fossils provide evidence for evolution

Key words

- ➤ selective breeding
- ➤ domesticated
- ➤ fossils

The modern theory of evolution by natural selection has developed over the past two centuries. It includes ideas from many different scientists. Their ideas were attempts to explain observations of the natural world. These observations included:

- how humans had produced new varieties of plants and animals by selective breeding
- fossils with similarities and differences to living species.

Selective breeding

Pet dogs (left) are the result of thousands of years of selective breeding. Humans selectively bred members of the Canidae family, which includes foxes, jackals, and wolves (right).

Some animals have characteristics that make them useful to humans. Some can do work for us. Some can be used as a source of food or materials, and we keep others as pets. Some plants are also useful, as sources of food and materials, and for decoration.

Throughout history, we have kept and bred animals and plants for our own benefit. We have deliberately selected individuals from each generation that have useful features. Sometimes, we have also selected individuals that lack undesirable features. We have allowed only these individuals to reproduce. In this way, we have tried to ensure that desirable features are passed on to the next generation of offspring, while undesirable features are not. This is **selective breeding**.

By mating individuals with different features, offspring with combinations of desirable features can be produced. However, sometimes desirable features are not part of the resulting combination in the offspring. It is not always possible to create the 'perfect' variety using selective breeding.

After many generations of selective breeding, the offspring can be quite different from their 'wild' ancestors.

The coastal strawberry produces large fruits, but they are white inside and cannot survive frosts.

The wild strawberry can survive frosts, but produces small fruits. The fruits are red outside and inside.

In the 18th century, French farmers bred coastal and wild strawberries to produce the garden strawberries we grow today. They can survive frosts and produce large fruits.

Garden strawberries are the result of selective breeding. They are hardy and produce large fruits. However, they are not as sweet as wild strawberries.

Feeding the world

In the Western world, most of what we eat comes from selectively bred plants and animals. Selective breeding has been essential in enabling us to feed the growing human population. For thousands of years, farmers have selected the crops that had the features they wanted, such as biggest yield and resistance to diseases.

Selective breeding has been used to create the common varieties of wheat, sheep, dogs, roses, and many other species. Animals that have been selectively bred to live and work with humans are said to have been **domesticated**. In some cases, the original 'wild' populations have become very rare or even extinct, replaced by the selectively bred varieties.

'Artificial selection' and natural selection

The famous naturalist Charles Darwin published a book called *On the Origin of Species* in 1859. In the first chapter of that book he discusses selective breeding – though he called it 'artificial selection'. He noted that selection by humans was a powerful driver of change in a species. In the following chapters he suggested that a similar process of selection in nature could explain the natural changes in species over time. Another naturalist called Alfred Russel Wallace proposed a similar explanation at around the same time. Their explanations of natural selection were the foundation of the modern theory of evolution.

Evidence from fossils

How do we know anything about organisms that lived tens of thousands to millions of years ago? All the evidence we have about them comes from **fossils** (and occasionally from organisms preserved in ice or resin). Fossils are the remains of dead organisms preserved in rocks.

Fossils can be compared to living organisms, and to other fossils, to reveal similarities and differences. Most fossilised organisms have similarities and differences to organisms alive today. This is evidence for the ideas that:

● organisms on Earth have changed over time

● new species evolve from earlier ones.

You may have seen cows out in the wild, but you've never seen a wild cow. The drawing shows a wild cow, known as an auroch. Modern cows were selectively bred from aurochs as early as 10 500 years ago. Aurochs are now extinct.

Can the principle of selection, which we have seen is so potent in the hands of man, apply in nature? I think we shall see that it can act most effectually.

A quote from Charles Darwin's book *On the Origin of Species*, published in 1859.

How much fossil evidence is there?

Scientists have collected millions of fossils. The huge amount of evidence – the **fossil record** – has helped to build our picture of evolution.

Conditions have to be just right for fossils to form. Very few organisms end up as fossils, so there are gaps in the fossil record. Sometimes a species does not seem to have a link to an earlier species.

Transitional species

Sometimes a fossil is found of an organism that shows some features of an older species and some of a newer species. These fossils are important evidence of evolution. They show one possible way that the newer species could have evolved from the older one. They are called **transitional species**.

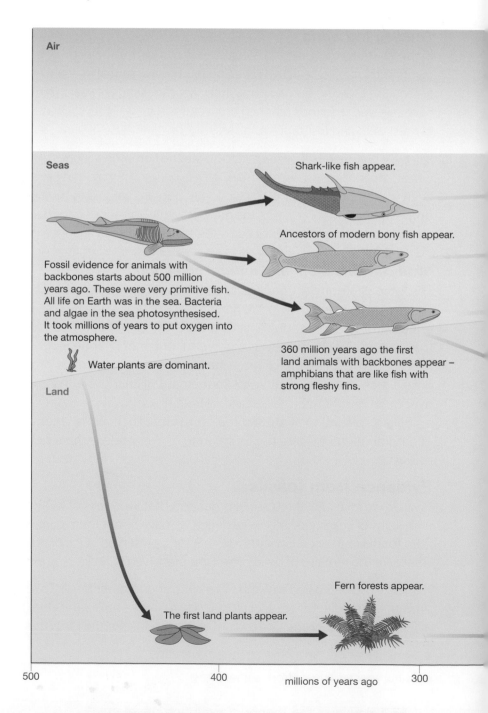

Air

Seas

Shark-like fish appear.

Ancestors of modern bony fish appear.

Fossil evidence for animals with backbones starts about 500 million years ago. These were very primitive fish. All life on Earth was in the sea. Bacteria and algae in the sea photosynthesised. It took millions of years to put oxygen into the atmosphere.

Water plants are dominant.

360 million years ago the first land animals with backbones appear – amphibians that are like fish with strong fleshy fins.

Land

Fern forests appear.

The first land plants appear.

500   400   millions of years ago   300

In the 19th century, fossils were found of a bird-like dinosaur called *Archaeopteryx*. It had feathers and broad wings, but also sharp teeth, claws with three fingers, and a long, bony tail. It was evidence that the first birds evolved from dinosaurs. Older fossils of small, meat-eating dinosaurs with feathers have since been found.

Common ancestors

You can see from the picture that later species have common ancestors. Mammals and dinosaurs both evolved from early reptiles. This idea suggests that all life on Earth started with one or a few common ancestors. Scientists agree that life on Earth probably began around 3500 million years ago, with a few very simple living organisms. These simple organisms evolved to produce the enormous variety of species on Earth today.

This *Archaeopteryx* fossil was discovered in Berlin in 1876. *Archaeopteryx* was about the same size as a modern crow.

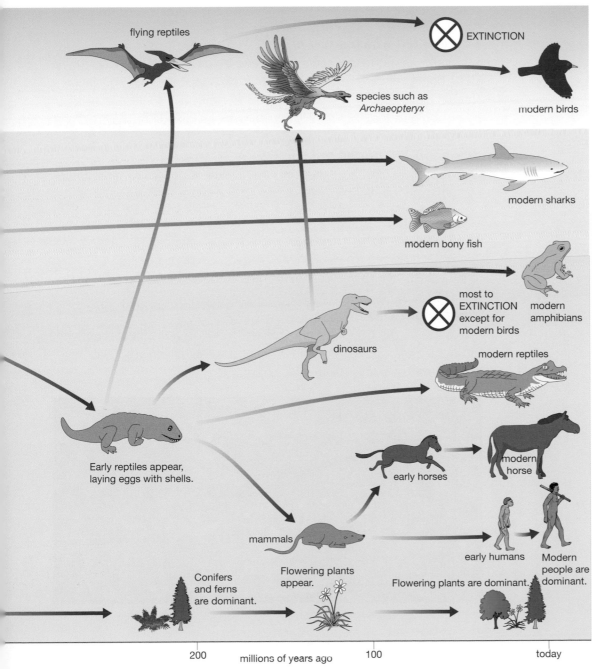

Dr Charlotte Brassey.

Secrets of the *Stegosaurus*

A fossil skeleton of a *Stegosaurus* dinosaur was put on display at the Natural History Museum, London, in 2014. Scientists at the museum had been studying the fossil to unlock its secrets.

Dr Charlotte Brassey was one of the scientists who worked on it. We asked her about what she's been able to learn about the *Stegosaurus* from its fossil.

What's special about this fossil?

This is the most complete *Stegosaurus* skeleton ever found. Scientists have known about this type of dinosaur for over 130 years, but the last big study was in 1914. Now we have lots of new technology and a much more complete skeleton. We can build new explanations about what the living animal looked like and how it lived.

Can you tell which sex it is?

Sometimes we find a dinosaur with an egg inside its body. Sometimes one of the bones contains a type of tissue that only appears during ovulation. These things would prove that it's female. But this *Stegosaurus* has neither. So it could be a male or a female that wasn't ovulating when it died.

What can the bones tell you?

Classification of dinosaurs into species is all based on the shapes of their bones – especially vertebrae (backbones) and skulls. This helps us trace the evolution of the various species.

We think this *Stegosaurus* died young. Its bones are different to other adult bones that have been found. Bones in the hips and neck aren't fused together fully. Also it's smaller than the other known examples of this species.

What about DNA?

DNA breaks down over time. After about 10 million years there is nothing left that we can use. This dinosaur lived about 150 million years ago. So the bones are the only evidence we have.

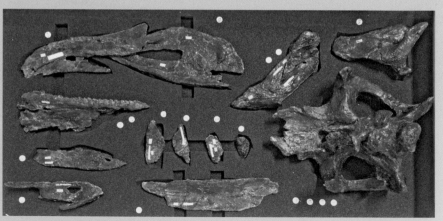

The skull of the *Stegosaurus* was found in many pieces, which had to be put back together.

How does modern technology help?

We've scanned every bone and made a three-dimensional model of the skeleton using a computer. Scientists around the world can study the *Stegosaurus* using the computer model. The real fossil is kept safely here at the museum.

We used the computer to add simulated muscles to the bones, and then made predictions about how the *Stegosaurus* moved. The computer tests thousands of combinations of movements, and selects the best combination.

We also used the computer model to predict the body mass of the *Stegosaurus* when it died. Our best estimate was around 1600 kg, which is about the same as a small rhino.

How do you know if these predictions are accurate?

We will never know. But we use the same computer models to make predictions about modern-day humans, elephants, and rhinos based on scans of their bones. When these predictions match the living animals, we assume the models are working well.

Stegosaurus is a very well-known dinosaur. But in reality very few *Stegosaurus* fossils have ever been found. Everything we know about the species comes from these few individuals. We have to make assumptions that the rest of the species resembled these individuals.

What else do you hope to find out?

Scientists have provided different explanations of the back plates on the *Stegosaurus*. Some think they were used as defence. Others think their large surface area was useful for temperature control. We will use a computer model to test the strength of the plates, to see how useful they would have been for defence. In the computer model we can apply forces to the plates until they break. We could never do this in real life because the fossil is far too important to break!

An artist's painting of what a living *Stegosaurus* might have looked like. But how do we know anything about animals that lived millions of years ago? We have to develop ideas from fossil evidence.

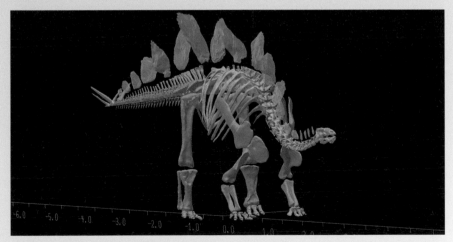

The computer model of the *Stegosaurus* skeleton.

B6.1 How was the theory of evolution developed?

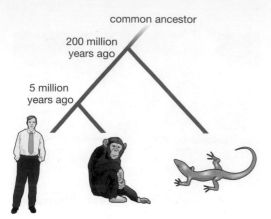

The idea that humans and apes evolved from a common ancestor was not popular at first.

The end of the story?

When Darwin and Wallace first published their ideas, many people in Victorian England disagreed with the idea of evolution by natural selection. They were unhappy about the idea that humans and apes shared a common ancestor. Some of Darwin and Wallace's ideas contradicted religious texts. Scientists debated the new ideas.

Since then, new evidence has come to light that supports the theory of evolution by natural selection:

- Fossils of transitional species have been found.

- Work on inheritance, DNA, and genomes explains how variation arises and is passed on.

- Modern examples of natural selection have been observed, for example, in head lice and bacteria.

Most scientists are now convinced that the evolution of species by natural selection is the best way to explain all of the available evidence.

Questions

1 Copy and complete the table below to compare selective breeding and natural selection.

Steps in selective breeding	Steps in natural selection
Living things in a species are not all the same.	Living things in a species are not all the same.
Humans choose the individuals with the feature that they want.	
These are the plants or animals that are allowed to breed.	
They pass their genes on to their offspring.	
More of the next generation will have the chosen feature.	
If people keep choosing the same feature, even more of the following generation will have it.	

2 Explain the impact of selective breeding on modern life.

3 Fossils provide evidence for evolution.
   a Write down two ways that fossils provide evidence for evolution.
   b What is the fossil record?

C: Evolution today

The explanation of evolution by natural selection has been modified over time as new evidence became available. For example, the modern theory of evolution includes the ideas that changes in DNA cause variation that can be inherited. Ideas about DNA and mutation were not developed until many decades after the deaths of Darwin and Wallace.

The explanation of evolution by natural selection is now described as a **theory**. A scientific theory is a general explanation that applies to a large number of situations or examples (perhaps to all possible ones). It has been tested and used successfully, and is widely accepted by scientists.

Find out about

● modern examples of evidence for evolution

Key word

➤ theory

Evidence from genomes

Over 98% of the DNA of humans and chimpanzees is the same. They were thought to share a common ancestor about 6 million years ago. They are both classified as primates.

Humans and mice are both mammals. About 85% of their DNA is the same. They shared a common ancestor about 75 million years ago.

Recent advances in gene technology have enabled scientists to compare the genomes of different organisms. There are many similarities and differences in the genomes of different species, just as there are in their body features.

These similarities and differences help us work out the evolutionary links between species. Two species that evolved from a common ancestor both inherited their genomes from it. This means they have many genetic variants in common with each other and with the ancestor. The more similar the genomes of two living things, the more closely related they are.

Modern examples of evolution
Head lice

People have been using poisons to kill head lice for many years. In the 1980s, doctors were sure that populations of head lice in the UK would soon be wiped out. But a few head lice survived the poisons. These lice bred and now parts of the country are fighting populations of 'superlice' that are resistant to the poisons.

Head lice are changing because of the actions of humans. But this wasn't selective breeding – we didn't choose resistant lice and breed them to produce 'superlice'. We introduced a poison into the lice's environment. This environmental change meant that some individuals survived while others did not. This was natural selection of the individuals best adapted to survive the poison.

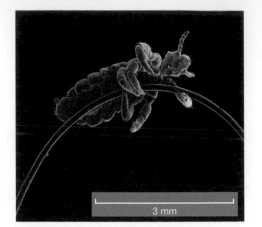

Head lice are quite common. They feed on blood.

1 For many years people used the same shampoo to kill head lice.

2 A few head lice in the population were able to survive. Their cells were probably able to break down the poison.

'Superlouse' was more likely to breed than the head lice killed by the poison.

3 Eggs laid by 'Superlouse' hatched into lice that also survived the poison.

4 These lice spread to other people and bred.

5 The number of resistant lice in the population increased. People couldn't get rid of their head lice.

6 Scientists developed a new poison to kill the head lice.

7 The cycle began again – and the species changed a little more.

Synoptic link

You can find out more about bacteria that are resistant to antibiotics in B2.5A *The best medicine*.

Antibiotic resistance

The spread of bacteria resistant to antibiotics is a big problem. Antibiotics are a relatively recent invention. The appearance of bacteria resistant to antibiotics is an example of rapid evolution.

- Genetic mutations happen at random, creating bacteria resistant to antibiotics.
- Use of antibiotics selects bacteria best adapted to survive them.

This example of evolution occurred very quickly, because bacteria reproduce quickly by dividing. Their DNA is copied every time they divide, and mutations most commonly arise when DNA is copied.

We cannot prevent the appearance of mutations that cause antibiotic resistance. But we can use antibiotics carefully to avoid selecting resistant individuals, and we can take measures to stop resistant bacteria spreading.

There are many other modern examples of evolution that you may wish to research, including changes in wildlife that live in cities, new species of cichlid fish in the African Great Lakes, and even the discovery of Nylon-eating bacteria!

Stopping the spread of antibiotic-resistant bacteria can be as simple as cleaning your hands.

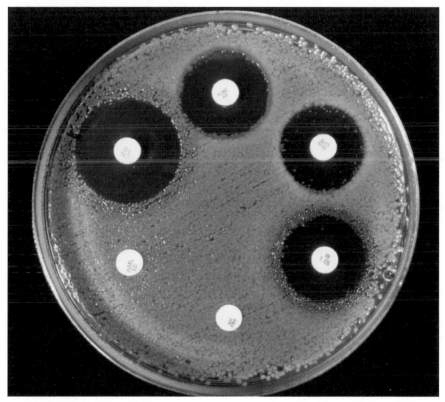

This Petri dish contains bacteria grown on agar jelly. Each white disc contains a different antibiotic. Two of the antibiotics have not killed the bacteria.

Questions

1 Explain why the appearance of antibiotic-resistant bacteria is an example of evolution.

2 Explain why evolution does not always take millions of years.

B6.2 How does DNA help us classify organisms?

A: Modern classification

Find out about

- classifying organisms
- the impact of DNA analysis on classification systems

Key words

➤ classification
➤ species
➤ genus
➤ kingdom
➤ archaea

Swedish biologist Carl Linnaeus classified many organisms in the 1700s. He gave each species a binomial (two-part) name. The first part is the genus and the second part is the species. The binomial name for this species of dolphin is *Delphinus delphis*. The potato plant is *Solanum tuberosum*.

There is an enormous diversity of living organisms on Earth. There are millions of known species, living in many different environments. Many species have yet to be discovered. Sorting organisms into groups helps us to understand the vast diversity of life on Earth.

Similarities and differences

The similarities and differences between organisms enable us to identify them and classify them into groups. Traditionally, identification and **classification** of organisms was based on:

- external and internal structures, organs, and tissues
- whether the organism can make its own food
- the environment in which the organism lives.

In the past 50 years or so, biologists have used additional evidence to classify organisms, including:

- cell structure
- DNA analysis.

This kind of evidence has only become available because new technologies enabled us to investigate organisms at the cellular and molecular level.

From species to kingdoms

As you know, a **species** is a group of organisms of the same type that can breed to produce fertile offspring. Several similar species can be grouped together into a **genus**. A number of similar genera (the plural of genus) can be grouped together, and so on. This creates a classification system of groups within larger groups. Larger groups contain more species with fewer characteristics in common.

Eventually, we reach several very large groups called **kingdoms**. One classification system that you may have used has five kingdoms. These are the plants, animals, fungi, protists, and monera (bacteria). Each kingdom contains millions of species.

A spider on a toadstool in the grass. How many kingdoms are represented in this photograph?

Biologists have debated for a long time how many kingdoms there should be. Different systems have been suggested over the years that included five, six, or even eight kingdoms. A common approach now is to divide the monera into two kingdoms – the bacteria and the **archaea** – making six kingdoms in total.

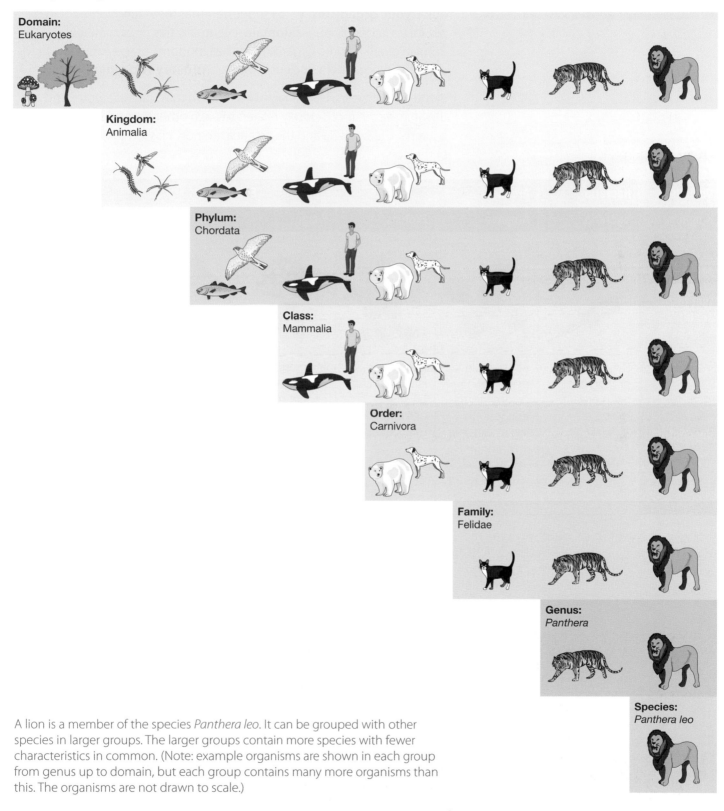

Domain:
Eukaryotes

Kingdom:
Animalia

Phylum:
Chordata

Class:
Mammalia

Order:
Carnivora

Family:
Felidae

Genus:
Panthera

Species:
Panthera leo

A lion is a member of the species *Panthera leo*. It can be grouped with other species in larger groups. The larger groups contain more species with fewer characteristics in common. (Note: example organisms are shown in each group from genus up to domain, but each group contains many more organisms than this. The organisms are not drawn to scale.)

From kingdoms to domains

The invention of the electron microscope in the 20th century enabled biologists to see the structures inside cells much more clearly. Organisms in the kingdoms of animals, plants, fungi, and protists have a visible nucleus containing genetic material. In other words, they are all **eukaryotic organisms**. It was suggested that these kingdoms could be put into a larger group – a **domain**. Eventually this became known as the domain of eukaryotes. Organisms in the kingdoms of the bacteria and the archaea do not have a nucleus – they are **prokaryotic organisms**.

In the past 40 years, DNA analysis has shown that the eukaryotes and the archaea are more closely related to each other than they are to the bacteria. The bacteria and the archaea are now classed as domains of their own.

Domain	Bacteria	Archaea	Eukaryotes			
Kingdom	Bacteria	Archaea	Protists	Fungi	Plants	Animals
example organisms (not to scale)						
description	Single-celled, prokaryotic microorganisms.	Single-celled, prokaryotic microorganisms, that tend to live in extreme environments.	Microorganisms that can be single-celled or multicellular.	Microorganisms that cannot make their own food. Most are multicellular, but some (yeasts) are single-celled.	Multicellular organisms that can make their own food.	Multicellular organisms that cannot make their own food.

Questions

1 Which kingdom are trees classified into?

2 Which domain are humans classified into?

3 Look at the diagram showing the classification of the lion. Suggest whether the lion is more closely related to tigers or polar bears, and explain why you think this.

4 What evidence was used to classify organisms into groups in the 19th century?

5 Carl Linnaeus classified many organisms in the 1700s. Suggest why a modern scientist may change the classification of an organism that was originally classified by Linnaeus.

A: Human impacts on biodiversity

The International Union for Conservation of Nature (IUCN) is an organisation that works to influence policy on the conservation of species and support on the ground conservation. They publish The IUCN Red List of Threatened Species™ which provides a measure of extinction risk. Species categorized as Vulnerable, Endangered or Critically Endangered are collectively described as 'threatened'.

Stories about endangered species, pollution, and the environment are often in the news. We're encouraged to live, use resources, and deal with waste in ways that do not harm the environment and the organisms living around us. Protecting ecosystems and the huge diversity of living organisms on Earth is important. All organisms, including humans, depend on other organisms and the environment for their survival.

However, human activities are causing a lot of damage to ecosystems. Our actions have resulted in many species becoming extinct, with other species heading the same way. Thankfully, there are things we can do to protect ecosystems and their communities of organisms. We can even help endangered species to recover from the threat of extinction.

What is biodiversity?

The enormous variety of different animals, plants, fungi, algae, and microorganisms on Earth is often referred to as the Earth's **biodiversity**. But there's more to it than that. The biodiversity of the Earth, or of a particular area, is the combination of:

- the diversity of living organisms
- the diversity of the organisms' genetic material
- the diversity of ecosystems in which the organisms live.

Are humans to blame for some extinctions?

Species become extinct when they cannot adapt to changes in their environment. New species are formed by natural selection. These processes happen naturally all the time. If new species are being formed, does it matter that some become extinct? Isn't extinction just part of life?

Find out about

- biodiversity
- how human activities can affect biodiversity in negative and positive ways
- the benefits and challenges of maintaining local and global biodiversity

Key word

➤ biodiversity

> If the Eiffel Tower were now representing the world's age, the skin of paint on the pinnacle-knob at its summit would represent man's share of that age; and anybody would perceive that that skin was what the tower was built for. I reckon they would. I dunno.

A quote from Mark Twain's essay *Was the World Made For Man?*, written in 1903. Did all life on Earth evolve just to support mankind? (And if mankind really is standing at the top of a tower of life, and we destroy that tower, what will happen to us?)

The mountain gorilla is one of the critically endangered species on the IUCN Red List.

The dodo was a species of bird. Humans caused the species to become extinct.

Over the past 200 years, the growth of industry and cities has eaten up the landscape, destroying natural ecosystems.

The problem is that the pattern of extinction today is different from what has been recorded in the past. The rate of species' extinction today is thousands of times higher than in the past. Scientists think that current extinctions are almost all due to human activity.

As dead as a dodo

In 1598, Dutch sailors arrived on the island of Mauritius in the Indian Ocean. In the wooded areas along the coast they found fat, flightless birds that they called dodos. By 1700, all the dodos were dead. The species had become extinct. The popular belief is that sailors ate them all. But this explanation appears too simple. Written reports from the time suggest that they were not very nice to eat.

Humans may have caused the extinction of the dodos without meaning to. The sailors brought with them rats, cats, and dogs. These animals may have attacked the dodos' chicks or eaten their eggs. The sailors also cut down trees to make space for their houses. This damaged and destroyed the dodos' habitat.

So human beings can cause other species to become extinct:

- directly, for example, by hunting
- indirectly, for example, by destroying habitat or bringing other species into an ecosystem.

The biodiversity of many areas is being reduced by human activities. These activities are related to increasing human population size, **industrialisation**, and **globalisation**. Many of our activities can result in ecosystems being damaged or destroyed, populations dying out, and species becoming extinct.

Humans now live all over the Earth, on almost every land mass. People, plants, animals, and other organisms are transported all over the Earth every day – from their natural habitats to places they were never supposed to be.

Once a species has become extinct, it cannot be brought back to life – it is gone forever. An ecosystem that has been damaged or destroyed may take a very long time to recover. Living sustainably and conserving biodiversity are important if the Earth is going to be a good home for future generations.

Human activities cause biodiversity loss by:

- changing, damaging, and destroying habitats
- removing or killing too many organisms
- introducing organisms and pathogens that do not belong in an ecosystem
- directly polluting and contaminating ecosystems
- polluting the atmosphere, which contributes to global climate change.

Key words

➤ industrialisation
➤ globalisation
➤ sustainability

Biodiversity and sustainability

One way to protect the Earth's biodiversity is to use its resources in a sustainable way. **Sustainability** means meeting the needs of people today without damaging the Earth for people of the future. One way to do this is to avoid using resources faster than they can be replaced.

The dangers of living unsustainably

Easter Island was once fertile and full of trees, but the people who lived there cut down the forest without ensuring that new trees grew. As you can see from the photo below, there are statues on Easter Island now, but trees are harder to find. The landscape is quite empty.

Stone statues (Moai) stand in an empty landscape on Easter Island.

How can we avoid the same fate as Easter Island? Most of us live in a 'take–make–dump' society. We take natural resources from our environment, make them into products, and then dump the waste. This system is not sustainable. It can only continue for a short time before things go wrong because:

- natural resources such as timber, clean air, fresh water, fertile soil, and fish stocks are being used more quickly than they are being replaced
- fossil fuels are being used unsustainably (as raw materials and fuel to make many of the products we buy)
- making and disposing of products creates a lot of waste
- waste can be harmful to many different organisms, and can stay in the environment for a long time.

Easter Island is located in the Pacific Ocean.

The benefits of biodiversity

Ecosystems that have high biodiversity tend to cope more easily with changes in conditions. In a drought, a community with lots of genetic variation is more likely to survive because some individuals will be better adapted to the drier conditions. Some species will not survive at all, becoming locally extinct. But if there are lots of different species in the ecosystem, some should survive.

Biodiversity and the survival of species are very important. Like all organisms, humans are part of the Earth's biodiversity and we depend on it for our survival. Ecosystems with high biodiversity provide us with many things that are useful. These are known as **ecosystem services**.

Most of us take it for granted that the Earth will continue to provide us with fresh water to drink, clean air to breathe, **fertile soil** for growing crops, and a supply of food such as fish and game. Ecosystem services include the different ways that living systems provide for human needs.

The growing human population is putting increasing pressure on ecosystem services worldwide. Understanding how ecosystem services work helps us to get the best out of them. This should help to avoid disasters similar to Easter Island, where ecosystem services broke down.

Foxgloves are very poisonous plants. But they have given us a powerful medicine to treat heart disease.

Underwater ecosystems are just as important as those on land. Coral reefs have been called 'the rainforests of the sea', and support thousands of species.

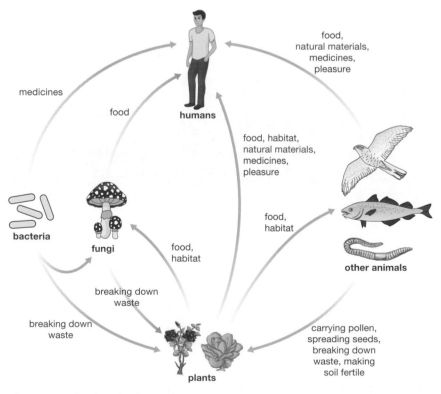

This is not a food web. The arrows show some of the different ecosystem services provided by different organisms. (Note: the organisms are not drawn to scale.)

Clean water

Mexico City has a population of around 20 million people. Rainfall comes between June and September and mostly falls on the surrounding mountains. There is a danger of flooding during the rains, but severe water shortages occur in the dry season. Mexico City has damaged the natural ecosystem by **deforestation** (cutting down the forest) and cattle grazing. Each year more water is used in the city than is replaced by rainfall.

Mexico City.

The forests on the mountains around Mexico City provide ecosystem services.

- The soil is rich in dead organic matter (DOM) from rotting leaves, and holds water like a sponge. The water gradually drains into the rivers supplying the city, so that there are reserves for the dry period.

- Tree roots reduce **soil erosion** by holding the soil together, and leafy branches prevent rain falling directly onto the soil. Rain drips slowly off the trees and is absorbed rather than running off the soil surface. Soil is not washed away. Eroded soil would silt up rivers and block drains.

- Water evaporation from the forest canopy generates clouds and rain, and cools the air.

Now the forests around Mexico City are being protected and restored. The people of Mexico City have recognised that damaging the ecosystem destroys a vital ecosystem service.

Year	Tiger numbers in India
1800s	45 000
1972	1827
1990s	3500
2010	1500

The number of tigers living in India has decreased dramatically over the past 200 years.

A story of conservation: the value of a tiger

Is a tiger worth more dead or alive? Sadly, living tigers are undervalued, but there is a huge market for dead tiger parts for traditional medicines. India is losing its tigers very rapidly. They may soon be extinct in the wild.

In 1972, the Indian Government established 28 tiger reserves spread across India. People were moved away and the vegetation was allowed to recover. Game increased rapidly and tiger numbers rose to 3500 by the 1990s.

In 1993, China banned the sale of farmed tiger products. More wild tigers were poached for traditional medicines. By 2008 tiger numbers in India had declined to just 1400.

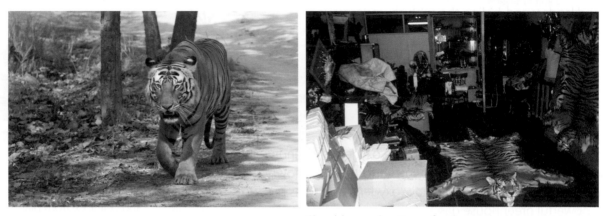

A Bengal tiger.

Should we value tigers for themselves and not just as a commodity?

The greatest good

Should people be forced to move away from tiger reserves? The buffer zones around the reserves make good grazing land for cattle and a source of wild game. There is little other good land for people to go to. But moving people away could benefit more people than it harms.

Why protect tigers?

Tigers are at the top of the food chain. A good tiger population means that the whole ecosystem is healthy. Tigers cannot survive without samba and chital (deer) prey. The deer will not survive without the right plants to eat.

The domestic chicken came from India. What other potential food sources do India's ecosystems hold?

Preserving India's ecosystems ensures that native plants and animals will survive. Wild plants provide medicines and food. Wild game can be harvested by humans for protein. Forests control climate and water resources.

Can we live without tigers? Tigers bring inspiration to many people; tourists come to see them; tiger reserves offer jobs to rangers and scientists. Without tigers, fewer tourists would come, and the whole ecosystem would change.

There are ethical reasons for conserving tigers and other wildlife. Many people believe that wildlife has a right to exist, for its own sake, without human interference. Some people think it is wrong to use tiger parts for medicines or to keep tigers in captivity.

Protecting biodiversity at different levels

1. Global level
Activities that contribute to global climate change, such as emissions of carbon dioxide, can be controlled.

Global warming and debates about carbon emissions are often in the news. Governments are encouraged to set tough targets to reduce emissions of greenhouse gasses.

2. Ecosystem level
Areas such as forests and coral reefs can be protected to reduce habitat destruction and pollution. This helps to protect many species.

The Murchison Falls National Park in Uganda protects 3480 km^2 of habitat for many species at risk from hunting, including Cape buffalo, elephants, lions, and leopards.

3. Species level
Individual species of organisms such as rhinos, elephants and whales can be protected to reduce overexploitation.

Some species, such as white and black rhinos, have been hunted almost to extinction because their body parts were seen as valuable commodities.

4. Genetic level
Gene banks can store genetic material from plants (seeds, pollen, tissues) and animals (sperm, eggs). Could this one day be used to resurrect a species that has become extinct?

Seeds, and the valuable DNA they contain, can be stored in a special facility, such as the Svalbard Global Seed Vault. The seeds should survive for hundreds or even thousands of years.

Questions

1 Dan thinks biodiversity is the range of organisms living on Earth. Explain why he is only partly correct.

2 Explain why industrialisation and globalisation are indirect causes of extinction.

3 Suggest three ecosystem services that are provided by trees.

4 Dead organic matter in the soil is decomposed by bacteria. The bacteria need oxygen for cellular respiration. Explain why earthworms are helpful to these decomposer bacteria.

5 Small animals in the ocean filter plastic granules from the water instead of their usual food.
   a Explain how this could affect populations of fish.
   b Suggest why the small animals cannot digest the plastic granules.
   c Suggest why packaging made of starch would be less harmful to ocean organisms.

6 Biodiversity can be protected at different levels.
   a Explain how it can be protected at global level, species level, and genetic level.
   b Suggest why protecting biodiversity at ecosystem level may be a better strategy than protecting an individual species.

Science explanations

B6 Life on Earth – past, present, and future

In this chapter you learnt about the link between DNA, variation, and evolution, and about evidence that supports the theory. You also learnt about the advantages and disadvantages of sexual and asexual reproduction, and how DNA evidence can be used to classify organisms. Finally, you learnt about how human activity can lead to loss of biodiversity, and ways in which we can protect it.

Fossils provide evidence for changes in species and the formation of new species over time. Fossils and other evidence led scientists to develop the theory of evolution by natural selection.

You should know:

- that there is usually extensive genetic variation within a population
- that one source of variation is genetic variants, which are caused by mutations
- that most mutations have no effect on an organism's phenotype, but those that do are usually harmful
- that very rarely a mutation creates a genetic variant that codes for a useful feature, leading to a phenotype better suited to the organism's environment
- that individuals and species compete for resources, and individuals better suited to compete are more likely to survive and reproduce
- that evolution is a change in the inherited characteristics of a population over a number of generations because of natural selection
- how evolution can lead to the formation of new species
- how selective breeding shows the effects that selection can have on a species
- that we rely on selectively bred plants and animals for food and other resources
- how the fossil record, including fossils of transitional species, provides evidence for evolution
- about modern examples of evidence for evolution, including antibiotic resistance in bacteria
- why most scientists accept the theory of evolution by natural selection
- how evidence from DNA can be used to help classify organisms
- what biodiversity is
- about negative and positive human interactions with ecosystems, and their impact on biodiversity
- about some of the benefits and challenges of maintaining local and global biodiversity.

Ideas about Science

The aim of science is to develop good explanations for observations of the natural world. The theory of evolution by natural selection was developed to explain the similarities and differences between living organisms and fossils. But scientific explanations do not emerge automatically from observations. Proposing an explanation involves creative thinking.

In the 1800s, Charles Darwin and Alfred Russel Wallace suggested that the evolution of species could be explained by natural selection. Their explanation was developed from many observations, using much creative thinking. It has been tested by many different scientists over the years, and has been modified and improved as new evidence became available. For example, observations of inheritance and DNA helped explain how helpful features could be passed to offspring if they were caused by changes in DNA.

The explanation is now regarded as a theory. A scientific theory is a general explanation that applies to a large number of situations or examples (perhaps to all possible ones), which has been tested and used successfully, and is widely accepted by scientists.

You should be able to:

- distinguish data from explanations and theories
- describe and explain examples of scientific explanations that were modified when new evidence became available.

Many human activities are damaging the environment, which can lead to loss of biodiversity. Scientists are coming up with ways to reduce the impact of our activities, and to protect species and the ecosystems they live in. One of the best things we can do is use natural resources sustainably. This means meeting our needs without damaging the Earth for future generations, often by using resources at the same rate as they can be replaced.

You should be able to:

- identify examples of unintended impacts of human activity on the environment
- describe ways that scientists work to reduce these impacts
- explain the idea of sustainability
- use data to compare the sustainability of alternative products or processes in specific situations.

Scientists are helping to come up with ways we can all use the Earth's resources sustainably.

B6 Review questions

1   Mutations can cause variation within the members of a species.

   a   Write down one choice from the list to complete each of the following sentences.

   a few   all   extensive   most   no   some

   i   There is usually .... genetic variation within a population.

   ii   .... mutations have no effect on an organism's phenotype.

   iii   .... mutations give the organism an advantage.

   b   What is a mutation?

   c   Explain how mutations cause variation within a population of a species.

2   Explain the link between competition and natural selection.

3   The chart shows the evolution of humans (*Homo sapiens*) over the past 7 million years, drawn using evidence from fossils.

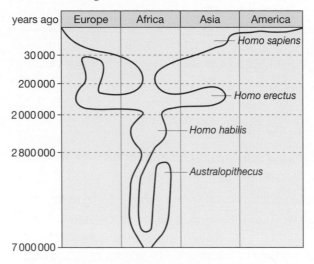

   a   Neanderthals are an extinct relative of humans. They did not evolve into *Homo sapiens*. Neanderthals became extinct just over 30 000 years ago. Which continent were Neanderthals mainly found in?

   b   Use the chart to answer these questions.

   i   Which of the statements below are true?

   A   All the species named on the chart evolved from a common ancestor.

   B   *Homo sapiens* appeared before *Homo erectus*.

   C   *Australopithecus* evolved from *Homo habilis*.

   D   *Homo habilis* spread to fewer continents than *Homo sapiens*.

   E   *Homo erectus* was mainly found in Africa.

   ii   Name one species on the chart that is not yet extinct.

4 Describe the impact that new technologies in the 20th century had on classification systems.

5 Describe three examples of negative human interactions with ecosystems that can cause biodiversity loss.

6 A scientist measured the levels of mercury in fish and birds living in a polluted bay. Small herbivorous fish had a concentration of 1 unit in their livers. Larger predatory fish had a concentration of 100 units, and herons that fed on the predatory fish had a concentration of 1000 units.

    a Give the term used to describe this change in concentration of pollutant in food chains.

    b Explain how the differences in mercury concentration happened.

7 The Food and Agriculture Organization (FAO) of the United Nations monitors commercially important populations of fish. It divides these populations into three categories: overfished, fully fished, and underfished. In a fully fished population, the fish are caught at the same rate as they are replaced by natural reproduction. The size of an overfished population can fall rapidly.

    a Which two categories represent sustainable levels of fishing?

    b The graph shows how the status of these fish populations changed between 1974 and 2010.

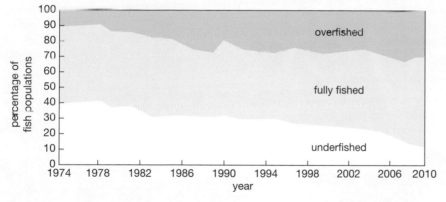

    i What percentage of fish populations were underfished in 1974?

    ii What percentage of fish populations were overfished in 1974?

    iii Which category's percentage changed the most between 1974 and 2010?

    iv Describe the overall trends shown in the graph.

    v Some people think we should increase the fishing of underfished populations, to help feed the growing human population. Suggest why this strategy would have to be carefully controlled.

    vi Suggest why climate change could cause the percentage of overfished populations to increase in the future.

B7 Ideas about Science

Why study Ideas about Science?

In order to make sense of the scientific ideas that you encounter in everyday life outside of school, you need to understand how science explanations are developed, the kinds of evidence and reasoning behind them, their strengths and limitations, and how far you can rely on them.

You also need to think about the impacts of science and technology on society, and how you respond to new ideas, artefacts, and processes that science makes possible.

What you already know

- Science explanations are based on evidence and as new evidence is gathered, explanations may change.
- How to plan and carry out scientific enquiries, choosing the most appropriate techniques and equipment.
- How to collect and analyse data and draw conclusions.

Case studies

In this chapter some of the ideas about science that you have studied across the course are explored in different contexts.

B7.1 Is there more to a playing field than grass and trees?
Find out about an investigation into the biodiversity of a playing field.

B7.2 Can science stop the spread of a deadly disease?
Find out how scientists in the UK helped to track down the cause of a deadly plant disease in Kenya, and whether science can help stop it.

Find out about

- investigating how different factors affect the plant species present in a playing field
- the accuracy of equipment and techniques used to measure abiotic factors
- sampling
- identifying trends and correlations in data
- sources of error and uncertainty

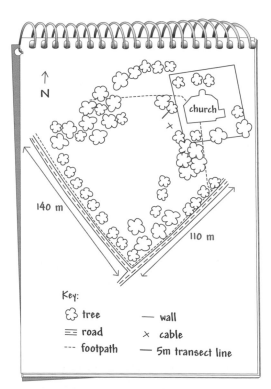

Key:
- 🌳 tree
- ≡≡ road
- --- footpath
- — wall
- × cable
- — 5m transect line

A sketch is a useful way to record where the samples were taken. It can also record key features of the site being investigated. A tape measure or trundle wheel should be used to measure the size of the area, to give a sense of scale.

Synoptic link

You can learn more about measuring biodiversity and abiotic factors in B3.4B *Investigating different habitats*, and also in B8 *Practical techniques*.

Every afternoon on his way home, Ali walks across a playing field next to a church. He had never paid much attention to the plants in the field. He thought it was just a big field of grass surrounded by trees. But one hot, sunny day he noticed something interesting.

The grass in the field was always mowed very short. But in one place an old cable meant that the grass could not be mowed. The grass around the cable was much taller, and there seemed to be a greater diversity of plants growing there. Ali also noticed that the grass had not been mowed underneath a nearby tree at the edge of the field. In that area, too, there were lots of different plants growing in the long grass.

He decided to investigate how the plant species changed with increasing distance from the tree.

Sampling

Ali decided to use quadrats to systematically sample the field at 10 places along a line transect.

- He placed a tape measure on the ground, with the zero end at the base of the tree.
- He laid out the rest of the tape measure, stretching away from the tree onto the mowed grass.
- He placed a 0.5 m × 0.5 m quadrat at the base of the tree. He made sure the bottom left corner of the quadrat was lined up with the 0 m graduation on the tape measure.
- He placed nine further quadrats at 1-metre intervals along the tape measure.

Ali used a tape measure to make a line transect stretching away from a tree. He placed quadrats at 1-metre intervals along the tape measure. The photograph shows quadrat number 3.

For each quadrat, Ali recorded:

● the percentage cover of bare soil

● the percentage cover of dead/brown grass

● the percentage cover of living/green grass

● the names of the plant species present

● the number of individual plants of each species.

He also measured the following abiotic factors for each quadrat:

● light intensity

● air temperature.

Finally, he collected a soil sample from each quadrat, and sealed it in an airtight container.

Look at the notes (right) about how Ali measured the air temperature. Do you think his measurements will be valid? Can you suggest a better way of taking the measurements?

> Light intensity was measured using a light meter placed on the ground in the bottom left square of each quadrat.
>
> Air temperature was measured using a thermometer held just above the ground in the bottom left square of each quadrat.
>
> A soil sample was dug up using a trowel from the bottom left square of each quadrat, after the species had been identified and recorded.

Ali made notes about how each abiotic factor was measured, and how the soil samples were collected.

The photo shows the equipment Ali used to measure the pH of each soil sample.

Drying soil samples in the oven.

Laboratory work

When he got back to the lab, Ali measured the pH and moisture content of each soil sample as follows.

Measuring soil sample pH

Ali weighed out 8 g of soil from the first soil sample into a small beaker. He added 5 cm^3 of distilled water and mixed it thoroughly with a glass rod to form a suspension. He then dipped a universal indicator test strip into the suspension, and compared the colour of the wet paper to a printed pH colour chart. He recorded the pH that was the nearest match to the colour of the paper. He then repeated the procedure for each of the other soil samples.

Measuring soil moisture content

Ali weighed a small foil tray and recorded its mass, and then transferred the remaining soil from the first sample into the tray. He then recorded the mass of the soil and tray together. He repeated these steps for the other nine soil samples. He placed all 10 trays of soil into the oven at 105 °C. He weighed the trays again after 3 hours, 6 hours, and finally 9 hours of drying.

He worked out what percentage of each soil sample's mass was water using the following equation:

$$\text{percentage of water} = \frac{(\text{total mass at 0 h} - \text{mass of tray}) - (\text{total mass at 9 h} - \text{mass of tray})}{(\text{total mass at 0 h} - \text{mass of tray})} \times 100\%$$

Sample	Mass of tray (g)	Mass of tray + soil (g) at 0 h	Mass of tray + soil (g) at 3 h	Mass of tray + soil (g) at 6 h	Mass of tray + soil (g) at 9 h
1	2.1	13.1	12.4	12.0	11.9
2	1.8	13.4	12.7	11.9	11.3
3	1.4	13.7	12.1	11.4	10.8
4	2.1	11.2	10.3	9.6	9.2
5	2.0	11.1	10.6	9.9	9.4
6	1.2	11.7	11.0	10.5	10.1
7	2.1	11.8	10.9	10.6	10.4
8	1.4	11.1	10.4	10.2	9.9
9	1.6	10.1	9.4	9.2	9.0
10	1.7	9.9	9.3	9.0	8.8

Ali recorded the mass of each soil sample and foil tray before putting them in the oven to dry, and then again after 3, 6, and 9 hours of drying.

Results

Here are the results from Ali's investigation.

Quadrat	1	2	3	4	5	6	7	8	9	10
Distance from tree (m)	0.00	1.00	2.00	3.00	4.00	5.00	6.00	7.00	8.00	9.00
Light intensity (W/m²)	127.4	126.0	167.0	206.8	303.2	673.2	567.8	1010.8	1071.0	1013.0
Air temperature (°C)	24.0 ± 0.5	24.0 ± 0.5	24.0 ± 0.5	27.0 ± 0.5	29.0 ± 0.5	31.0 ± 0.5	25.0 ± 0.5	28.0 ± 0.5	29.0 ± 0.5	30.0 ± 0.5
Soil pH	6	6	6	6	6	6	6	6	6	6
% water	10.9	18.1	23.6	22.0	18.7	15.2	14.4	12.4	12.9	13.4
Bare soil (% cover)	16	0	4	0	0	0	0	0	8	12
Dead/brown grass (% cover)	0	0	52	0	8	0	0	8	24	16
Living/green grass (% cover)	80	100	44	100	92	100	100	92	68	72
Ivy (squares)	7									
Creeping buttercup (squares)	17									
Yellow archangel (squares)	5	2								
Lords-and-ladies (squares)	1									
Stinging nettle (squares)	1	3								
Ground ivy (squares)		4								
Rough hawkbit (squares)			1	1		1				
Common chickweed (squares)				3	4					
Greater plantain (squares)							1			
White clover (squares)							11	8	7	2

Lords-and-ladies (red fruits; *Arum maculatum*) and ivy (dark green leaves; *Hedera helix*).

Creeping buttercup (*Ranunculus repens*).

Ali's analysis

This is what Ali wrote about his measurements of light intensity and air temperature:

> There is a clear trend in the data for light intensity. Also, there seems to be a correlation between the air temperature and the light intensity.

However, he was concerned about the results for quadrat 7. Here is what he wrote:

> A large cloud moved across the sky when I took the measurements for quadrat 7. I measured the light intensity in this quadrat again when it was sunny, and the reading was 791.4 W/m².

He compared his measurements of soil pH and percentage of water with some **secondary data**. He found the secondary data on a website.

> I found some secondary data about soil on the UK Soil Observatory website (www.ukso.org). The website said that soil in this area is 'freely draining and slightly acidic'. My results support that statement.

Ali noted that the grass had been mowed starting from 1.78 m away from the tree. Closer to the tree, the grass has not been mowed. He wrote the following conclusion about the different plant species present:

> The species found in quadrats 1 and 2 are different from those found in quadrats 3–10. Quadrats 3–10 were on ground that had been mowed. The changes in species present must be because of the mowing.

Stinging nettle (*Urtica dioica*).

White clover (*Trifolium repens*).

Questions

1 **a** In the analysis, Ali wrote that a large cloud may have affected the measurements from quadrat 7. Which measurements are likely to have been affected?

   b Explain why it would be better to measure the light intensity in each quadrat at different times during the day and calculate the mean of the measurements.

   c Look back at Ali's note about how he measured the light intensity in each quadrat. Apart from cloud cover, state two other factors that could affect the measurement of light intensity in a quadrat and describe how Ali controlled them.

2 Ali used universal indicator test strips and a colour chart to measure the pH of the soil sample suspensions. Write down one piece of equipment Ali could have used to measure the pH more accurately.

3 Ali used a thermometer to measure the air temperature. Suggest whether each of the following is a source of systematic error, random error, or a mistake when taking measurements using the thermometer.

   a Reading the wrong number from the scale on the thermometer.

   b Recording the reading to the nearest graduation.

   c The thermometer is only accurate to $\pm 0.5\ ^\circ C$.

4 Ali dried the soil samples in the oven for 9 hours. He measured the mass of each sample at 0 h, 3 h, 6 h, and 9 h. Should he have continued to dry the samples for longer than 9 h? Explain your answer.

5 **a** Describe the trend in light intensity as the distance from the tree increases. How would you explain this trend?

   b Is there a correlation between light intensity and air temperature? Explain your answer.

6 Ali suggests that his results support the statement that the soil is 'freely draining and slightly acidic'. Do you agree? You should evaluate the evidence from Ali's results in your answer.

7 Ali noticed some other plant species growing in the field, such as a creeping thistle. However, there were no thistles in any of the quadrats, so the species was not recorded in his results. Explain why his quadrats are not a representative sample of the whole field.

8 Do you agree with Ali's conclusion that 'the changes in species present must be because of the mowing'? Explain your answer.

Find out about

- how applications of science can help to identify the cause of a disease and make a positive difference to people's lives
- risks and benefits in the context of seeds and plant disease
- government regulation and ethical issues
- why it is helpful for scientists to communicate their work to a range of audiences

In September 2011, farmers in Kenya in Africa reported that their maize crops were dying of a mysterious disease. Soon the disease had become an epidemic, and was headline news in Kenyan newspapers. Farmers' livelihoods were being destroyed, and many people faced hunger.

The disease became known as maize lethal necrosis (MLN) disease. Nobody knew what was causing it, or how to stop it spreading. A year later, Kenya's *Daily Nation* newspaper reported that the disease was still puzzling scientists.

An international collaboration was formed between scientists from Kenya and the *Fera Science Ltd.* (Fera) in the UK, to try to identify the cause of the disease.

The importance of maize

Maize is often called corn. Most people in the UK have eaten cornflakes, sweetcorn, popcorn, and corn-on-the-cob. In the UK, these foods are luxuries – people don't depend on them to survive.

An outbreak of MLN disease was reported in Kenya in 2011. Since then it has spread to Tanzania, Uganda, Rwanda, and Ethiopia – a total area twelve times larger than the UK.

Ears of maize, also called corn.

In many countries, including Kenya, maize is a staple food. It is commonly grown by **subsistence farming** – the farmers grow enough maize to feed themselves and their families, and sell any they have left. Maize is a source of food, flour, and oil. Many people in Kenya depend on maize for their survival.

On the left is a field of healthy maize. In the field on the right, all the maize has been killed by MLN disease.

Key word

- ➤ subsistence farming
- ➤ diagnostic test
- ➤ genome sequencing

Maize lethal necrosis disease

In 2012, farmer Joseph Koech told Kenya's *Daily Nation* newspaper: 'Every farmer has been affected by the disease. The leaves turn yellow and then the whole plant rots away within weeks. It continues to spread by the day.'

The symptoms of MLN disease: the maize plant leaves turn yellow, the plant stops growing properly and can become bent over, and the ears do not fill properly with seeds.

MLN disease destroys up to 100% of the maize in an infected area. Because nobody knew what was causing the disease, farmers and officials had no idea how to stop it spreading. Farmers did not know what to do with infected plant material. Many fields with infected crop were abandoned.

Identifying the cause

An important first step in stopping the spread of a disease is identifying the pathogen that is causing it. But this is not always easy.

It was known that many different bacteria, fungi, and viruses could cause symptoms similar to those seen in the infected maize. For most of these pathogens, **diagnostic tests** had been developed by scientists around the world. The tests use microscopy, monoclonal antibodies, and gene probes to detect the presence of particular pathogens. The problem is deciding which tests to use when you don't know which pathogen you're looking for. Scientists at Fera ran several tests on samples of the infected maize, but the results did not provide a clear diagnosis of the pathogen.

Instead of continuing to test for specific pathogens, the team decided to try a new approach. They used **genome sequencing** to provide information about all of the microorganisms present in the infected maize.

Scientists from Fera travelled to Kenya, where local people showed them the devastated crops.

A scientist from Fera looks at insects caught in a maize field in Kenya. Could one of these insect species be involved in spreading the disease?

Dr Julian Smith (left) was one of the scientists from Fera who helped to identify the cause of MLN disease.

Many farmers in Kenya keep some of the seeds from each harvest, and re-plant them to grow the next year's crop. This can be a problem when the seeds are infected with a pathogen.

Key word

➤ correlation

Genome sequencing

The team at Fera sequenced pieces of genetic material from all of the organisms found in several samples of infected maize. The results suggested that there were over 100 different microorganisms present in the samples, including bacteria, fungi, and viruses. The challenge was to work out which one was the cause of MLN disease. Dr Adrian Fox of Fera summarised the problem: 'Just because you find a pathogen in the sample, this doesn't mean it's the cause of the disease.'

A crop can be infected with multiple pathogens but not develop disease. The crop may be resistant, or the pathogens may not be present in high enough numbers. These pathogens are a 'background population'. Fera needed to distinguish the disease-causing pathogen from the background population.

They sequenced genetic material from samples of maize with symptoms, and compared these to sequences from samples of maize without symptoms. The results showed that two viruses – maize chlorotic mottle virus (MCMV) and sugarcane mosaic virus (SMV) – were always present in maize with symptoms, and usually absent from maize without symptoms. The team then used diagnostic tests to look for the two viruses in hundreds of samples of maize with and without symptoms. Again, the two viruses were always found in diseased samples and usually not found in healthy ones. This was evidence of a **correlation** between the symptoms and the pathogens.

Stopping the spread

MLN is not a new disease. It had been reported previously in South East Asia and North America. So why had it now appeared in East Africa?

Tests showed that both of the viruses must be present for the symptoms of MLN disease to appear. SMV has been widespread in East Africa for some time and is spread by insects. Fera's investigation showed that the outbreak of the disease began when MCMV was introduced to the region. MCMV was probably introduced in infected seed imported from Southeast Asia. At a regional level, the trade of MCMV-infected seed within and between countries led to the epidemic of MLN disease.

In 2015 the disease was still affecting large parts of Kenya, and had spread to neighbouring countries. The *Star* newspaper in Kenya reported that maize production had decreased by 10 million bags in 2015 – almost a quarter of the total annual production.

So what can be done to stop the spread of the disease? The traditional approach is to use selective breeding to create resistant varieties of the crop, but this can take years. Many subsistence farmers cannot afford to wait that long. In the meantime, some other useful strategies include:

- a programme of seed testing
- the sale and use of only certified seed. This is seed that has been tested and shown to be free of MCMV.

Can science solve the problem?

It has been difficult to implement the widespread sale and use of certified seed in East African countries. The idea has to be supported by governments, but they have been slow to introduce regulations. These regulations would control the testing of seed for certification, and could ban the sale of uncertified seed.

Many farmers save seed from their own crop and replant it. This means they don't have to buy new seed. They also sell surplus seed to make money, but this can spread the disease if the seed is infected.

Companies such as *Kenya Seed* have been producing and selling certified maize seed, but it is more expensive than uncertified seed. Lack of government funding and increasing cost of agricultural inputs threaten to push the price even higher.

Should subsistence farmers be forced to pay for certified seed, or should they be allowed to continue using cheaper, uncertified seed in the traditional way? This question cannot be answered using science alone. As Dr Julian Smith from Fera suggests, 'We can share our test results and new technologies, but the people in Kenya must decide what happens next.'

Education programmes like this one is West Africa can help local farmers to recognise the symptoms of plant diseases and understand ways of stopping the spread.

Questions

1 Applications of science can make a positive difference to people's lives. Describe one application of science that could help farmers in Kenya threatened by MLN disease.

2 Globalisation has led to an increase in international travel, trade, and cooperation.
   a International trade is a useful human activity, but it can have unintended negative impacts on the environment. The outbreak of MLN disease in Kenya could be one example of a negative impact of international trade. What is the evidence for this in the case study?
   b Explain how globalisation was helpful in tackling the MLN disease epidemic.

3 A subsistence farmer buys some maize seed from her neighbour, rather than buying seed from the company *Kenya Seed*. Describe the benefits and risks of this course of action. You should identify who benefits and who is at risk.

4 The government of an East African country is debating whether or not to introduce a new regulation. The regulation would make it compulsory for famers to buy and use only certified maize seed, and illegal to buy, sell, or use uncertified seed. Different people hold different views on this ethical issue.
   Summarise the view that may be held by:
   a a company that sells certified seed
   b a scientist working to stop the spread of MLN disease
   c a subsistence maize farmer.

5 Using MLN disease as an example, explain why scientists must communicate their work to a range of audiences.

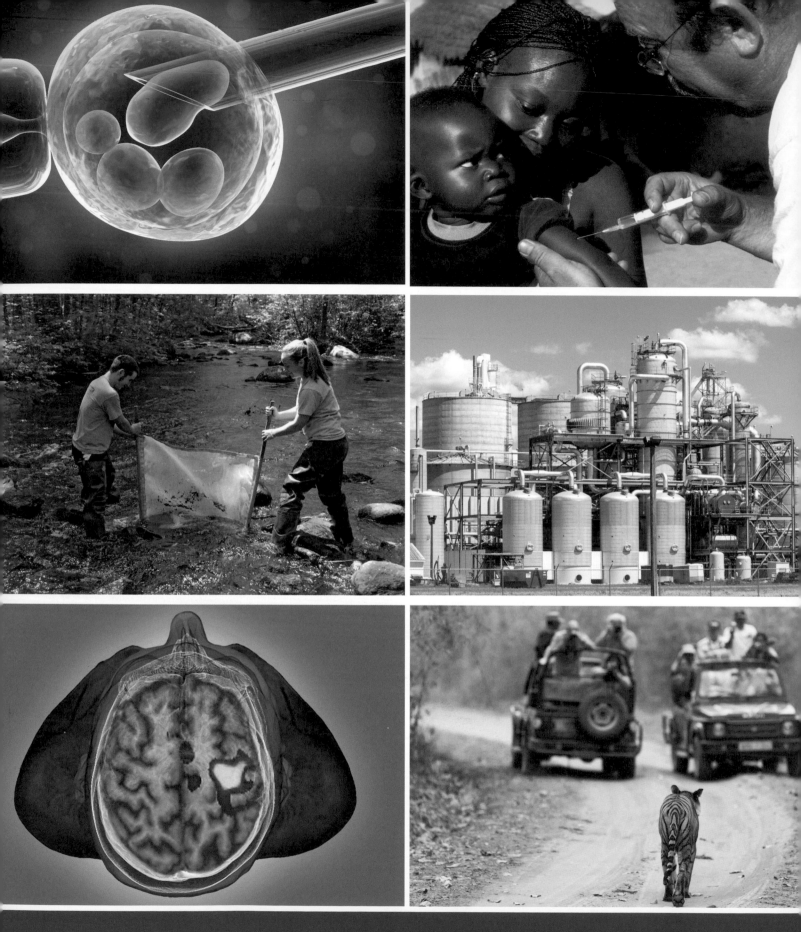

Ideas about Science

Learning about the Ideas about Science in this course will help you understand how scientific knowledge is obtained, how to respond to science stories and issues in the wider world, and the impacts of scientific knowledge on society.

IaS1: What needs to be considered when investigating a phenomenon scientifically?

The aim of science is to develop good explanations for natural phenomena. There is no single 'scientific method' that leads to explanations, but scientists do have characteristic ways of working. In particular, scientific explanations are based on a cycle of collecting and analysing data.

Usually, developing an explanation begins with proposing a hypothesis. A hypothesis is a tentative explanation for an observed phenomenon ('this happens because…').

The hypothesis is used to make a prediction about how, in a particular experimental context, a change in a factor will affect the outcome. A prediction can be presented in a variety of ways, for example, in words or as a sketch graph.

In order to test a prediction and the hypothesis upon which it is based, it is necessary to plan an experiment that enables data to be collected in a safe, accurate, and repeatable way.

In a given context you should be able to:

- use your scientific knowledge and understanding to develop and justify a hypothesis and prediction

- suggest appropriate apparatus, materials, and techniques, justifying the choice with reference to the precision, accuracy, and validity of the data that will be collected

- use scientific quantities (such as mass, volume, and temperature) and know how they are measured

- identify factors that need to be controlled and how they could be controlled

- suggest an appropriate sample size and/or range of values to be measured, and justify the suggestion

- plan experiments and describe procedures to make observations, collect data, and test a prediction or hypothesis

- identify hazards associated with the data collection and suggest ways of minimising the risk.

Can you plan an experiment to safely collect data on the plant species growing in a field? What equipment would you use, and how? What are the risks, and how can they be minimised?

IaS2: What processes are needed to draw conclusions from data?

The cycle of collecting, presenting, and analysing data usually involves translating data from one form to another, mathematical processing, graphical display, and analysis; only then can we begin to draw conclusions.

A set of repeat measurements can be processed to calculate a range within which the true value probably lies and to give a best estimate of the value (mean).

Displaying data graphically can help to show trends or patterns, and to assess the spread of repeated measurements.

Mathematical comparisons between results and statistical methods can help with further analysis.

When working with data you should be able to:

- produce appropriate tables, graphs, and charts to display the data
- use the appropriate units and be able to convert between units
- use prefixes (from tera to nano) and powers of 10 to show orders of magnitude
- use an appropriate number of significant figures.

When displaying data graphically you should be able to:

- select an appropriate graphical form, using appropriate axes and scales
- plot data points correctly, drawing an appropriate line of best fit and indicating uncertainty (e.g., range bars)

When analysing data you should be able to:

- identify patterns or trends
- use statistics (range and mean)
- obtain values from a line on a graph (including gradient, interpolation, and extrapolation).

Number of cigarettes smoked per day	Number of cases of cancer per 100 000 men
0–5	15
6–10	40
11–15	65
16–20	145
21–25	160
26–30	300
31–35	360
36–40	415

This table shows data on cigarette smoking and lung cancer in men. What would be the best way to display the data graphically?

Experiments and data must be evaluated before we can make conclusions based on the results. There could be many reasons why the quality (accuracy, precision, repeatability, and reproducibility) of the data could be questioned and a number of ways in which it could be improved.

Data can never be relied on completely because observations may be incorrect and all measurements are subject to uncertainty, arising from the limitations of the measuring equipment and the person using it.

A result that appears to be an outlier should be treated as data, unless there is a reason to reject it (e.g., measurement or recording error).

In a given context you should be able to:

- discuss the accuracy, precision, repeatability, and reproducibility of a set of data
- identify random and systematic errors that are sources of uncertainty in measurements
- explain the decision to discard or retain an outlier
- suggest improvements to an experiment and explain why they would increase the quality of the data collected
- suggest further investigations that could be done.

A prediction is based on a tentative explanation (a hypothesis). When collected data agree with the prediction, it increases our confidence that the explanation is correct. But it does not *prove* that the explanation is correct. Disagreement between the data and the prediction indicates that one or other is wrong, and decreases our confidence in the explanation.

In a given context you should be able to:

- use observations and data to make a conclusion
- explain how much the data increases or decreases confidence in a prediction or hypothesis.

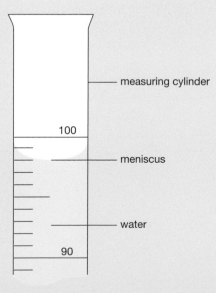

The volume of a liquid can be measured using a measuring cylinder. But there will be uncertainty in the measurement. Can you suggest a source of random error, a source of systematic error, and a mistake that could be made when taking this kind of measurement?

IaS3: How are scientific explanations developed?

Scientists often look for patterns in data to identify correlations that can suggest cause-and-effect links. They then try to explain these links.

The first step is to identify a correlation between a factor and an outcome. The factor may be the cause, or one of the causes, of the outcome. In many situations, a factor may not always lead to the outcome, but increases the chance (or the risk) of it happening. In order to claim that the factor causes the outcome we need to identify a process or mechanism that might account for how it does this.

You should be able to use ideas about correlation and cause to:

- identify a correlation in data presented as text, in a table, or as a graph
- distinguish between a correlation and a cause-and-effect link
- suggest factors that might increase the chance of a particular outcome in a given situation, but do not always lead to it
- explain why individual cases do not provide convincing evidence for or against a correlation
- explain why you would accept or reject a claim that a factor is a cause of an outcome, based on the presence or absence of a causal mechanism.

Scientific explanations and theories do not 'emerge' automatically from data, and are separate from the data. Proposing an explanation involves creative thinking. Collecting sufficient data from which to develop an explanation often relies on technological developments that enable new observations to be made.

As more evidence becomes available, a hypothesis may be modified and may eventually become an accepted explanation or theory.

A scientific theory is a general explanation that applies to a large number of situations or examples (perhaps to all possible ones), which has been tested and used successfully, and is widely accepted by scientists. A scientific explanation of a specific event or phenomenon is often based on applying a scientific theory to the situation in question.

You should be able to:

- describe and explain examples of scientific explanations that have developed over time, and how they were modified when new evidence became available.

Explanations about species that lived millions of years ago are developed from evidence provided by fossils. How did fossils of *Archaeopteryx* help scientists to build explanations about the evolution of birds?

Findings reported by an individual scientist or group are carefully checked by the scientific community before being accepted as scientific knowledge. Scientists are usually sceptical about claims based on results that cannot be reproduced by anyone else, and about unexpected findings until they have been repeated (by themselves) or reproduced (by someone else).

Two (or more) scientists may legitimately draw different conclusions about the same data. A scientist's personal background, experience, or interests may influence their judgements.

An accepted scientific explanation is rarely abandoned just because new data disagrees with it. It usually survives until a better explanation is available.

You should be able to:

● describe the 'peer review' process, in which new scientific claims are evaluated by other scientists

Models are used in science to help explain ideas and to test explanations. A model identifies features of a system and rules by which the features interact. It can be used to predict possible outcomes. Representational models use physical analogies or spatial representations to help visualise scientific explanations and mechanisms. Descriptive models are used to explain phenomena. Mathematical models use patterns in data of past events, along with known scientific relationships, to predict behaviour; often the calculations are complex and can be done more quickly by computer.

Models can be used to investigate phenomena quickly and without ethical and practical limitations, but their usefulness is limited by how accurately the model represents the real world.

For a variety of given models (including representational, descriptive, mathematical, computational, and spatial models) you should be able to:

● use the model to explain a scientific idea, solve a problem, or make a prediction

● identify limitations of the model.

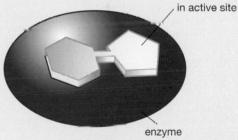
substrate molecule in active site

enzyme

How does the lock-and-key model help us to explain how enzymes work?

IaS4: How do science and technology impact society?

Science and technology provide people with many things that they value, and that enhance their quality of life. However, some applications of science can have unintended and undesirable impacts on the quality of life or the environment. Scientists can devise ways of reducing these impacts and of using natural resources in a sustainable way.

You should be able to:

● describe and explain examples of applications of science that have made significant positive differences to people's lives.

Everything we do carries a certain risk of accident or harm. New technologies and processes can introduce new risks. The size of a risk can be assessed by estimating its chance of occurring in a large sample, over a given period of time.

To make a decision about a course of action, we need to take account of both the risks and benefits to the different individuals or groups involved. People are generally more willing to accept the risk associated with something they choose to do than something that is imposed, and to accept risks that have short-lived effects rather than long-lasting ones. People's perception of the size of a particular risk may be different from the statistically estimated risk. People tend to overestimate the risk of unfamiliar things (like flying as compared with cycling), and of things whose effect is invisible or long-term (like ionising radiation).

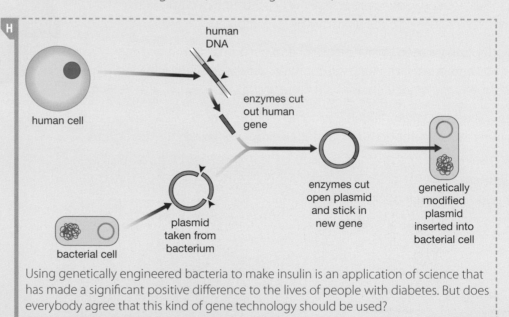

Using genetically engineered bacteria to make insulin is an application of science that has made a significant positive difference to the lives of people with diabetes. But does everybody agree that this kind of gene technology should be used?

You should be able to:

- identify examples of risks that have arisen from a new scientific or technological advance

- for a given situation:
  - identify risks and benefits to the different individuals and groups involved
  - discuss a course of action, taking account of who benefits and who takes the risks
  - suggest reasons for people's willingness to accept the risk
  - **H** distinguish between perceived and calculated risk.

Some forms of scientific research, and some applications of scientific knowledge, have ethical implications. In discussions of ethical issues, a common argument is that the right decision is one which leads to the best outcome for the greatest number of people.

Where an ethical issue is involved you should be able to:

- suggest reasons why different decisions on the same issue might be appropriate in view of differences in a personal, social, economic, or environmental context

- make a decision and justify it by evaluating the evidence and arguments

- distinguish questions that could be answered using a scientific approach from those that could not

- state clearly what the issue is and summarise different views that may be held.

Scientists must communicate their work to a range of audiences, including the public, other scientists, and politicians, in ways that can be understood. This enables decision-making based on information about risks, benefits, costs, and ethical issues.

You should be able to explain why scientists should communicate their work to a range of audiences.

Animal testing helps us to discover new medicines. How does this benefit humans, and what are the ethical issues?

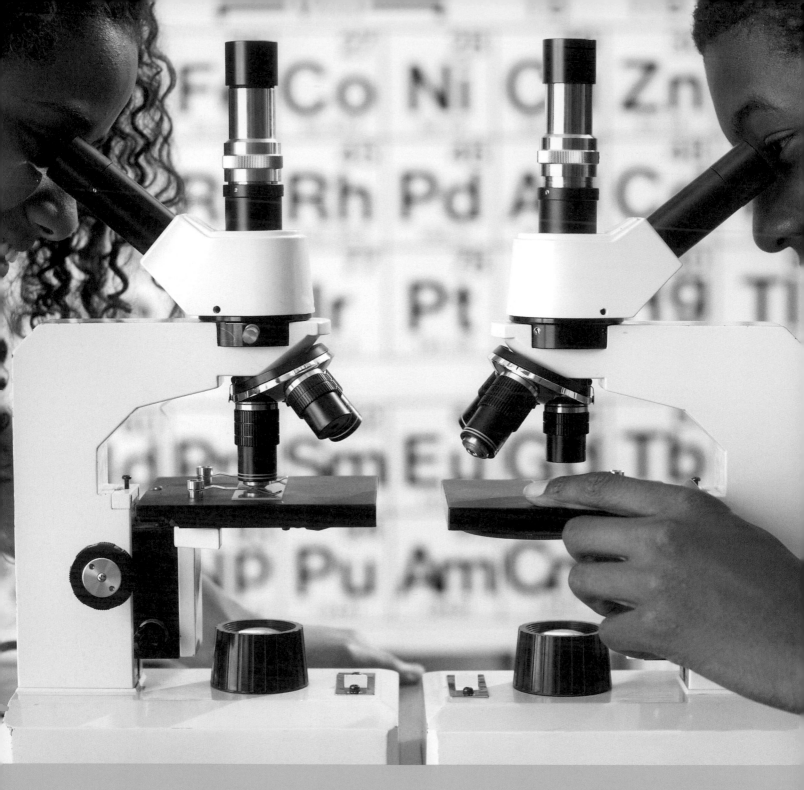

B8 Practical techniques

Why study practical techniques?

Practical work is an essential part of science. It helps us investigate what happens inside our own bodies and in the world around us. It also helps us explain how and why these things happen.

The aim of science is to develop good explanations for observations of the natural world. Scientific explanations are based on data, which must be collected.

Practising practical techniques helps us to collect data in a safe, ethical, and repeatable way, and to improve the accuracy of the data we collect. It is important that different people use the same techniques and standard procedures to collect data. This means data can be compared more easily.

Practical work not only helps us to develop explanations, but also to test explanations proposed by others. Understanding some of the ways scientists collect data helps us to make informed decisions about scientific issues in the news and in our own lives.

Practical techniques

- Using appropriate apparatus to make and record a range of measurements accurately, including length, mass, time, temperature, volume of liquids and gases, and pH.

- Safely using appropriate heating devices and techniques, including a Bunsen burner and a water bath.

- Using appropriate apparatus and techniques to observe and measure biological changes and processes.

- Measuring physiological functions and responses to the environment in living plants and animals, safely and ethically.

- Measuring rates of reaction by monitoring the production of gas, uptake of water, and colour change of an indicator.

- Using appropriate sampling techniques to investigate the distribution and abundance of organisms in an ecosystem, and measuring biotic and abiotic factors.

- Using appropriate apparatus, techniques, and magnification, including microscopes, to make observations of biological specimens.

A: Measuring length, temperature, and volume

Find out about

● using appropriate apparatus to accurately measure and record length, temperature, and volume of liquids and gases

Apparatus and materials

➤ ruler
➤ thermometer
➤ measuring cylinder
➤ syringe
➤ pipette
➤ analogue meter

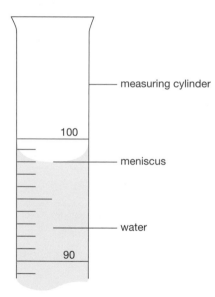

Liquid in a measuring cylinder has a curved top surface. This is called the **meniscus**. The volume is read from the bottom of the meniscus. You should always make the reading at eye level. The volume of water in the diagram is 98 ± 0.5 cm³.

Key words

➤ graduation
➤ uncertainty
➤ meniscus

Many measuring instruments have a linear scale. This is a series of equally spaced lines, or **graduations**. You use the scale to read off a value.

You may have to use a linear scale on:

● a ruler (to measure length)
● a thermometer (to measure temperature)
● a measuring cylinder, syringe, or pipette (to measure the volume of a liquid or gas)
● an analogue meter.

When a measurement falls between two graduations, it has to be estimated. This means there will be **uncertainty** in the measurement.

Procedure

1   Look at the scale you are reading. Usually, not all of the graduations will be marked with a number. Decide what the distance between two graduations represents.

2   If the reading is between two graduations, decide which graduation it is closest to. Record this as the measurement.

3   Record the uncertainty of the measurement as ± half the smallest graduation.

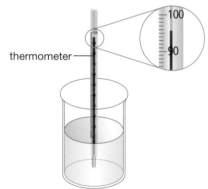

A thermometer has a linear scale. The coloured liquid inside the thermometer expands as the temperature goes up, and rises up the narrow glass tube. Here, the reading is between 95 °C and 96 °C, but is closest to 96 °C. We would record the measurement as 96 °C ± 0.5 °C.

B: Measuring mass

Mass is measured using a balance. Choose a balance with a level of accuracy fit for the purpose of the task – to how many significant figures do you need to measure the mass?

If a very sensitive balance is used, it is necessary to shield the balance from drafts to get an accurate measurement.

Find out about

- using appropriate apparatus to accurately measure and record mass

Apparatus and materials

- ➤ balance
- ➤ weighing vessel

Procedures

Weighing direct

1 Check that the balance is clean and reading zero.

2 Place a suitable empty weighing vessel on the balance platform. Set the display to zero. (This is known as taring the balance.)

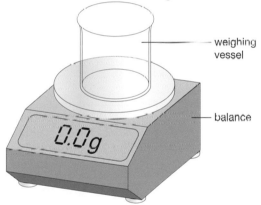

weighing vessel

balance

3 Place the sample in the weighing vessel on the balance platform.

4 The reading on the balance is the mass of the sample.

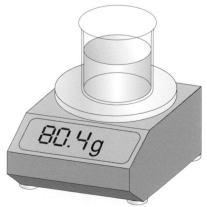

Weighing by difference

1 Check that the balance is clean and reading zero.

2 Place the sample in a suitable clean, dry weighing vessel on the balance platform. Record the mass.

3 Transfer the sample to another container. Weigh the weighing vessel and record its mass.

4 Calculate the mass of the sample from the difference between the two measurements.

C: Measuring time

Find out about

● using appropriate apparatus to accurately measure and record time

Apparatus and materials

➤ timer (such as a stopwatch or stop clock)

Time is measured using a timer such as a stopwatch or stop clock.

You may wish to measure, for example, how long it takes for a photosynthesising plant to release a particular volume of carbon dioxide gas, or for an organism to respond to a stimulus.

Procedure

A stopwatch showing a time of 42 s. This is an example of a digital meter.

A stop clock showing a time of 42 s. This is an example of an analogue meter.

1 At the starting point, start the timer (or if the timer is already running, make a note of the start time).

2 At the end point, stop the timer (or if the timer needs to keep running, make a note of the end time).

3 If the start time was not zero, subtract the start time from the end time.

Uncertainty

A digital stopwatch may show the time to tenths (0.1) or hundredths (0.01) of a second. However, when using a stopwatch, human reaction time can be up to 0.5 s. Therefore, you may need to record your measurement of the time with an uncertainty of ±0.5 s.

Minutes and seconds

Remember that 1 minute is divided into 60 seconds. This means that 1 minute 50 seconds is *not* the same as 1.50 minutes. (1.50 minutes is one-and-a-half minutes, which is 1 minute 30 seconds.)

D: Measuring pH

Universal indicator solution contains a mixture of compounds that change colour at different levels of acidity or alkalinity. When universal indicator solution is added to a sample, the colour it turns shows the pH of the sample.

Alternatively, you can use pH test strips and match them to a colour chart, or use a digital pH meter.

Procedure for measuring pH by colour matching

1   Put about 5 cm³ of the sample solution into a test tube.

2   Add three drops of universal indicator solution, and shake the tube gently from side to side to mix. Or dip a pH test strip into the solution.

3   Use the colour chart to estimate the pH value of the sample solution. Compare the colour of the solution or the pH test strip with the chart and decide the nearest match.

pH value

red
orange
yellow
green
blue
indigo
violet

1
2
3
4
5
6
7
8
9
10
11
12
13
14

CHECK SAFETY
Never work
unsupervised

wear eye
protection

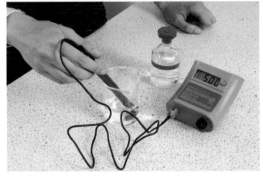

An alternative way to measure pH is to use a digital pH meter. The probe of the pH meter must be cleaned using distilled water between each reading. The probe is very delicate so must be handled with care and not used to stir the solution.

E: Heating substances and controlling their temperature

Find out about

- safely using appropriate heating devices, including a Bunsen burner and a water bath

Apparatus and materials

- ➤ tripod
- ➤ gauze
- ➤ container
- ➤ Bunsen burner
- ➤ thermometer
- ➤ clamp, boss, and stand
- ➤ water bath
- ➤ heat-proof mat

CHECK SAFETY
Never work
unsupervised

**wear eye
protection**

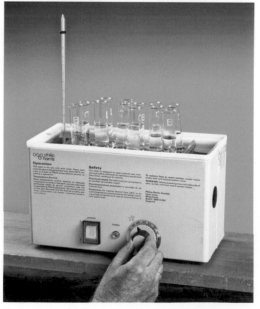

The water in an electric water bath is kept at a constant temperature. The temperature is set using the dial, and is controlled automatically by an electric heater and thermostat. It can be checked using the thermometer. Test tubes containing a substance can be placed in the water to heat the solution and help control its temperature.

Substances can be heated using a Bunsen burner. The substance is placed in a glass or ceramic container on a tripod and gauze.

It can be difficult to keep a substance at a constant temperature using a Bunsen burner. In this case it is helpful to use a water bath. You may use an electric water bath, or you may make a water bath by immersing the sample container in a large beaker of water at the desired temperature. The temperature of a water bath cannot be raised above the boiling point of water (100 °C).

Procedure for heating with a Bunsen burner

1 Place the tripod and gauze on a heat-proof mat. Ensure that the tripod is stable and will not tip over. Place the container of substance on top.

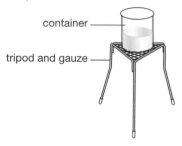

container

tripod and gauze

2 Turn the ring on the Bunsen burner so that the air hole is completely closed. This will produce a yellow safety flame when the Bunsen burner is lit. Turn on the gas, and then light the Bunsen burner.

Bunsen burner

3 Turn the ring on the Bunsen burner so that the air hole is open. The hottest flame (a blue roaring flame) will be produced when the air hole is completely open. Place the Bunsen burner under the tripod.

4 Use a thermometer to measure the temperature of the substance. Hold the thermometer so that the bulb is in the centre of the substance. The bulb should not touch the bottom or sides of the container. Use a clamp to hold the thermometer in place to reduce the risk of spillage when working with higher temperatures, and when monitoring the temperature over a length of time.

F: Comparing rates of reactions

We can compare the rate of a chemical reaction in different conditions. For example, is it faster at higher temperature? Or we can compare the rates of two different reactions in the same conditions.

A simple way to do this is to compare how long it takes for each reaction to reach an end point. A reaction that reaches the end point more quickly has a faster rate.

The rate of each reaction could be calculated using the following equation:

$$\text{rate of reaction (/s)} = \frac{1}{\text{time taken to reach end point (s)}}$$

Procedures

Timing how long it takes for an indicator to change colour

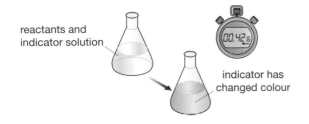

reactants and indicator solution

indicator has changed colour

Mix the reactants in a flask with an indicator solution, and start the timer. Stop the timer when the indicator has completely changed colour.

Timing how long it takes for a solution to turn cloudy (for reactions that form a precipitate)

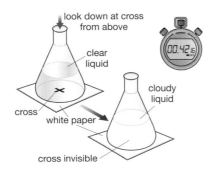

look down at cross from above

clear liquid

cloudy liquid

cross

white paper

cross invisible

Mix the liquids in a flask and start the timer. Stop the timer when you can no longer see the cross.

Timing how long it takes for a solid reactant to dissolve

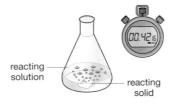

reacting solution

reacting solid

Mix the solid and liquid in a flask and start the timer. Stop the timer when you can no longer see any solid.

Find out about

● measuring rates of reaction by monitoring colour change of an indicator

● using appropriate apparatus and techniques to observe and measure biological changes and processes

Apparatus and materials

➤ conical flask
➤ timer
➤ indicator solution
➤ white card with black cross on it

CHECK SAFETY
Never work unsupervised

wear eye protection

G: Measuring the rate of a reaction

Find out about

- measuring rates of reaction by monitoring the production of gas
- using appropriate apparatus to make and record measurements accurately, including volume of gases
- using appropriate apparatus and techniques to observe and measure biological changes and processes

Apparatus and materials

- ➤ conical flask
- ➤ bung
- ➤ delivery tube
- ➤ tray of water
- ➤ measuring cylinder
- ➤ clamp, boss, and stand
- ➤ gas syringe
- ➤ balance
- ➤ cotton wool
- ➤ timer

CHECK SAFETY
Never work
unsupervised

wear eye
protection

We can measure the rate of a chemical reaction by finding the quantity of product produced or the quantity of reactant used up in a fixed time.

Procedures

Collecting gas in a measuring cylinder or gas syringe

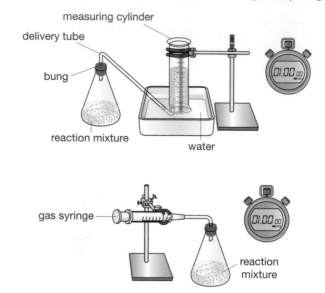

1 Record the volume of gas in the measuring cylinder or gas syringe at the starting point of the reaction.

2 Record the volume of gas in the measuring cylinder or gas syringe 60 seconds later or at the end point of the reaction.

The rate of reaction is calculated using the following equation:

$$\text{rate (cm}^3\text{/s)} = \frac{\text{change in volume of gas (cm}^3)}{\text{time taken (s)}}$$

Measuring the loss of mass as a gas is formed

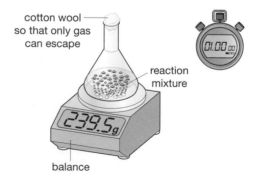

1 Record the mass at the starting point of the reaction.

2 Record the mass 60 seconds later or at the end point of the reaction.

The rate of reaction is calculated using the following equation:

$$\text{rate (g/s)} = \frac{\text{change in mass (g)}}{\text{time taken (s)}}$$

H: Measuring physiological functions and responses

We can measure physiological functions in living organisms. These functions include, for example, pulse rate, body temperature, and blood pressure. This usually means the organism experiences close interaction and physical contact with the measuring equipment and its operator.

We can also measure the ways that living organisms respond to changes in their environment. It is sometimes easier to observe and measure these responses in controlled conditions in a laboratory. But this means removing organisms from their natural environment.

Procedure

Practical work that involves living organisms must be done safely and ethically.

When measuring a living organism:

- make sure that you and the organism are safe at all times
- minimise the risk of harm to you and the organism
- minimise physical contact, handling, and discomfort
- wash your hands before and after handling the organism
- return the organism to its natural environment as soon as possible.

When the organism is a person, you must also:

- explain to them what you will do before you start
- make sure you have their permission to take the measurement before you start.

Example: measuring pulse rate

The diagram in the margin shows some places where you can feel the pulse. The pulse can be measured from the radial artery in the wrist. The person should be relaxed.

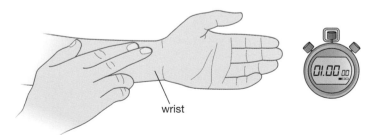

wrist

1 Place your first and second fingers along the radial artery and press gently against the bone. Apply enough pressure to feel the pulse but not so much that the artery is blocked.

2 Count the number of pulses in 1 minute. This gives the pulse rate in beats per minute.

Find out about

- measuring physiological functions and responses to the environment in living plants and animals, safely and ethically

Apparatus and materials

➤ timer

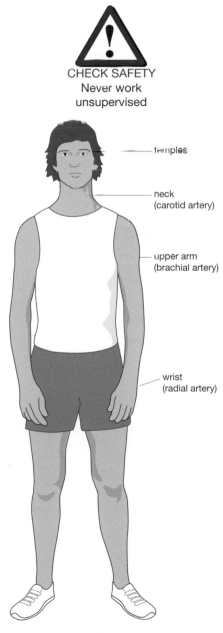

CHECK SAFETY
Never work
unsupervised

temples

neck
(carotid artery)

upper arm
(brachial artery)

wrist
(radial artery)

Major pulse points of the upper body.

I: Investigating the distribution and abundance of organisms

Find out about

- using appropriate sampling techniques to investigate the distribution and abundance of organisms in an ecosystem
- measuring biotic factors

Apparatus and materials

- quadrat
- tape measures
- two sets of numbered cards
- random number table
- species identification key
- net
- sample containers

CHECK SAFETY
Never work
unsupervised

Wash your hands after outside activities, especially after being in water.

Synoptic link

You can learn more about why these techniques are useful in B3.4B *Investigating different habitats.*

Fieldwork often involves investigating:

- the distribution of organisms within an ecosystem (where they are found)
- the abundance of organisms within an ecosystem (the population size of each species).

We must always consider the effects that the fieldwork will have on the ecosystem being studied, and work carefully to minimise damage.

Procedures

Using random quadrats to measure abundance

It could take a long time to count every individual organism in a population, particularly in a large area. It is quicker to count the organisms in a sample of the population – for example, in a small area.

To get a representative sample of an area such as a field, place quadrats randomly, as follows:

1 Mark out two edges of the area using tape measures.

2 Choose two numbers at random (from two sets of numbered cards or a random number table). Place the bottom left corner of the quadrat at these coordinates, for example 11,7.

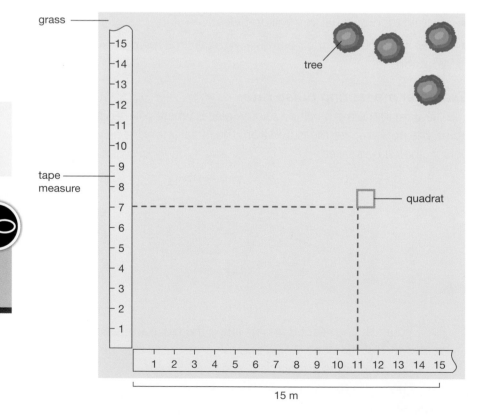

3 Repeat step 2 to generate coordinates for further quadrats. The more quadrats you use, the bigger the sample. A bigger sample is more likely to be representative of the whole area.

For each quadrat:

4 Identify the plant species inside the quadrat. You may want to use a species identification key.

5 Count and record the number of individual plants. Count any individual that is more than halfway inside the quadrat. Ignore any individual that is less than halfway inside the quadrat.

species	number of individuals
daisy	3

6 For grasses and bare soil, count how many squares inside the quadrat they appear in. Count squares that are half filled or more than half filled. Ignore squares that are less than half filled.

To estimate the percentage cover, first work out what percentage of the quadrat area is represented by each square. In the quadrat shown in the margin there are 25 squares, so each square represents 4% of the quadrat area. To estimate the percentage cover, multiply the number of squares by the percentage represented by each square.

	number of squares	percentage cover
grasses	22	88
bare soil	3	12

Once you have recorded data for all quadrats in the sample, the data can be processed to work out the mean number of individuals of each species in the sample, and the mean percentage cover of grasses and bare soil. If the sample is representative, these values can be used to make conclusions about the whole area (see B3.1B).

Using a transect to measure distribution and abundance

Quadrats can be placed at regular intervals along a straight line called a transect. This is done to investigate how the types of plants change gradually from one area to another (for example, from the edge of a wood into an open field) as well as to measure their abundance in each quadrat.

The plant species present in each quadrat and their numbers (or percentage cover) are recorded in the same way as for random quadrats.

Kick sampling a river

Kick sampling is a way to sample indicator species of animals in a river or stream.

1 Hold a net in the flowing water.

2 Kick the river bed upstream of the net for at least 30 seconds.

3 Tip the contents of the net into a sample container such as a white plastic tray.

4 Identify and count any animals present in the sample. Return the organisms to the river as soon as possible.

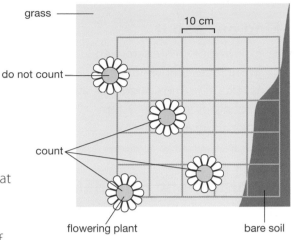

Individual plants that are more than halfway inside the quadrat are counted. Individual plants that are less than halfway inside the quadrat are not counted.

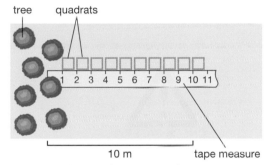

Using quadrats with a transect line.

Kick sampling a river.

J: Measuring abiotic factors in an ecosystem

Find out about

- using appropriate sampling techniques to investigate the distribution and abundance of organisms in an ecosystem
- measuring abiotic factors

Apparatus and materials

- ➤ light meter
- ➤ thermometer
- ➤ sample containers with screw-on lids
- ➤ balance
- ➤ oven
- ➤ beaker
- ➤ pH meter
- ➤ universal indicator
- ➤ pH test strips

CHECK SAFETY
Never work
unsupervised

Wash your hands after outside activities, especially after being in water.

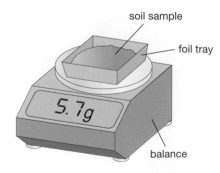

soil sample

foil tray

5.7g

balance

The distribution and abundance of organisms within an ecosystem is affected by abiotic factors, including:

- light intensity
- temperature
- water content of the soil
- soil pH.

light meter

1003 W/m²

Procedures

Measuring light intensity

Use a light meter. Take the reading at ground level or at the same height above the ground each time. Make sure not to cast your own shadow over the light meter.

Measuring temperature

The temperature of air, soil, and water can be measured using a thermometer.

When measuring soil temperature, first make a hole in the soil that is just wide enough to put the thermometer into. Do not try to push the thermometer into soil without first making a hole. Measure the temperature at the same depth each time.

Measuring soil water content

1   Collect soil samples at the same depth each time.
2   Put each sample in a container and screw tightly shut to reduce the loss of water from the soil by evaporation. Label the container clearly.
3   In the laboratory, weigh each sample and record its mass.
4   Transfer each sample to a foil tray (one sample per tray). Heat the trays of soil in the oven at 105 °C to dry the soil. Carefully weigh the samples after 12 hours, then return them to the oven. Record the mass of each sample every 3 hours, until it does not change for three consecutive weighings.

Use the following equation to work out the percentage of each soil sample's mass that was water:

$$\text{percentage of water} = \frac{\text{mass before drying (g)} - \text{mass after drying (g)}}{\text{mass before drying (g)}} \times 100$$

Measuring soil pH

1   Put the soil sample in a beaker.
2   Add distilled water and stir to make a suspension.
3   Filter the suspension to remove clumps of soil, stones, and dead organic matter.
4   Use a pH meter to measure the pH of the filtrate.

K: Measuring the rate of water uptake by a plant

Plants lose water because of transpiration and photosynthesis. They take in an equal amount of water to replace it. So the amount of water they take in is affected by the rates of transpiration and photosynthesis.

We can measure the rate of water uptake by a plant using a simple potometer.

Procedure

1 Fill the sink with water.

2 Hold the bottom end of the shoot under the water. Cut off the end of the shoot. Leave the shoot in the water.

3 Insert the bottom end of the pipette into one end of the flexible tubing.

4 Make sure the tubing fits tightly over the end of the pipette, then cover the join with waterproof jelly to make a watertight seal.

5 Hold the pipette and tubing under water in the sink until they are filled with water.

6 Carefully insert the cut end of the shoot into the empty end of the tubing.

7 Remove the shoot, tubing, and pipette from the water and clamp into position as shown in the diagram.

8 Make sure the tubing fits tightly over the end of the shoot and then cover the join with waterproof jelly.

9 Carefully dry the leaves of the shoot.

10 Wait until the meniscus of the water in the pipette has moved down to the next graduation.

11 Record the volume of water in the pipette, then start the timer.

12 After at least 30 minutes, stop the timer. Record the time and the volume of water in the pipette.

13 Calculate the rate of water uptake using the following equation:

$$\text{rate of water uptake (cm}^3\text{/s)} = \frac{\text{change in volume of water (cm}^3\text{)}}{\text{time taken for volume to change (s)}}$$

Find out about

● measuring rates of reaction by monitoring the uptake of water
● safely measuring physiological functions and responses to the environment in a living plant

Apparatus and materials

➤ sink
➤ water
➤ leafy plant shoot
➤ scissors, knife, or blade
➤ graduated pipette
➤ flexible tubing
➤ waterproof jelly
➤ clamp, boss, and stand
➤ timer

CHECK SAFETY
Never work unsupervised

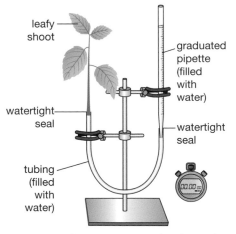

A potometer that you use in school may be different to this one. Make sure you know how to use it before you start.

L: Using a light microscope

Find out about

- using appropriate apparatus, techniques, and magnification, including microscopes, to make observations of biological specimens

Apparatus and materials

➤ light microscope
➤ specimen on microscope slide

CHECK SAFETY
Never work
unsupervised

wear eye
protection

Using a light microscope enables us to observe biological specimens at the cellular level.

Procedure

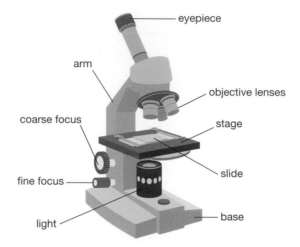

eyepiece

arm

objective lenses

coarse focus

stage

fine focus

slide

light

base

1   Switch on the light so that it shines light through the specimen.

2   Turn the smallest objective lens towards the stage. Rotate the turret until you hear a click.

3   Use the coarse focus to move the lens to its lowest position.

4   Clip the slide on the stage.

5   Look down the eyepiece lens with one eye. Try to keep the other eye open.

6   Adjust the coarse focus to move the lens away from the specimen until the image is as clear as possible.

7   Now turn the medium or high objective lens towards the stage.

8   Adjust the fine focus to move the lens away from the specimen until the image is as clear as possible.

> Never look down a microscope without a slide on the stage.
> Always start with the objective lens near the slide and move it away so that you do not smash the slide.

M: Preparing a temporary slide

A sample viewed under a light microscope must be thin enough to allow light to pass through it. Living organisms, therefore, must be very small to be seen under a microscope.

Non-living material can be squashed or sliced thinly. Dyes can be used to stain the sample and show features that cannot otherwise be seen.

Procedure

1 Place the specimen on a microscope slide.

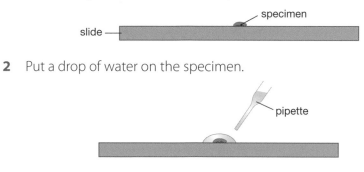

2 Put a drop of water on the specimen.

3 Use a mounted needle to lower a coverslip over the drop of water slowly. Avoid air bubbles.

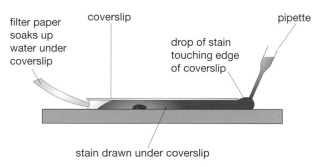

4 Examine the specimen with a light microscope.

5 If required, stain the specimen by placing a drop of stain touching the edge of the coverslip. Use a piece of filter paper to soak up water from the opposite edge of the coverslip. It should draw the stain under the coverslip.

filter paper soaks up water under coverslip

coverslip

drop of stain touching edge of coverslip

pipette

stain drawn under coverslip

6 Re-examine the slide under the microscope.

Find out about

● using appropriate apparatus, techniques, and magnification, including microscopes, to make observations of biological specimens

Apparatus and materials

➤ light microscope
➤ specimen
➤ microscope slide
➤ pipette
➤ water
➤ mounted needle
➤ coverslip
➤ stain
➤ filter paper

CHECK SAFETY
Never work unsupervised

wear eye protection

N: Uncertainty in measurements

Find out about

- uncertainty in measurements
- systematic and random errors as sources of uncertainty

Key words

➤ variation
➤ true value
➤ uncertainty
➤ random error
➤ systematic error
➤ outlier
➤ accuracy
➤ precision

If you measured the pulse rate of 20 people, you wouldn't get the same value for everybody. There would be **variation** in the data – in other words, there would be a range of measured values. This is because there is variation in the population. Everybody is different, and there are factors (such as diet, exercise, and stress) that cause the pulse rate to be different in each person.

If each of the 20 people measured the length of the same piece of paper, you might expect them all to come up with the same length. But you could still see variation in the data. This is due to the measurements themselves. Measured values are usually different to the **true value**, and each time a measurement is taken we can't be certain how close it is to the true value. So it is useful to give an indication of **uncertainty** when recording a measurement.

For example, you could record the volume of a liquid as 15.30 ± 0.05 cm^3. From this we can see that:

- the measured value (which could be the mean of several measurements) is 15.30 cm^3

- there is uncertainty in the measurement

- the person who took the measurement is confident that the true value is between 15.25 cm^3 and 15.35 cm^3.

Sources of uncertainty

There are two general sources of uncertainty in measurements: systematic errors and random errors.

Random error causes repeated measurements to give different values. This can happen, for example, when making judgements about the colour change at an end point or when estimating the reading from a thermometer.

A random error is not a mistake made by the person taking the measurement. A random error is a source of variation in measurements that cannot be eliminated, although there are often things we can do to reduce the amount of variation it causes.

Systematic error causes all repeated measurements to be the same amount higher or lower than the true value. This can happen when using an incorrectly calibrated measuring instrument or when taking measurements at a consistent, but wrong, temperature.

An example of how random and systematic errors can occur

The two different types of error are illustrated in the following example. The flask shown in the margin is used to measure out 25 cm^3 of a solution.

For the flask in the diagram, the manufacturer states that the line indicates a measured volume of 25.00 ± 0.06 cm^3. So even if the measured volume of liquid is exactly aligned with the marked line each time, the volume could be consistently larger or smaller than 25.00 cm^3 (by up to 0.06 cm^3). This is systematic error.

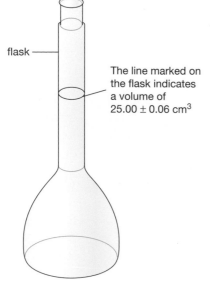

flask

The line marked on the flask indicates a volume of 25.00 ± 0.06 cm^3

This flask is used to measure a particular volume of liquid.

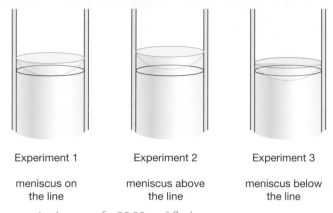

Experiment 1	Experiment 2	Experiment 3
meniscus on the line	meniscus above the line	meniscus below the line

Random errors in the use of a 25.00 cm³ flask.

It is difficult to fill a flask with liquid so that the bottom of the meniscus is aligned exactly with the marked line. In experiments 2 and 3 in the diagram above, the meniscus is not aligned exactly – so the measured amounts are not 25.00 cm³. This is an example of random error. We could reduce the size of the error by aligning the meniscus as closely as possible to the line each time, but it would be very difficult to completely eliminate this source of error.

The difference between 'error' and 'mistake'

Random and systematic errors are not mistakes. Mistakes are failures by the person taking the measurement, such as taking readings from a sensitive balance in a draught. Mistakes of this kind lead to **outliers** in results, and should be avoided by people doing practical work. If it is known that a mistake was made when taking or recording a measurement, the measurement should be taken again.

Accuracy and precision

An analysis or test is often repeated to give a number of measured values, which are then averaged to produce the result.

● **Accuracy** describes how close a result is to the true or 'actual' value.

● **Precision** is a measure of the spread of measured values. A big spread indicates a greater uncertainty than a small spread.

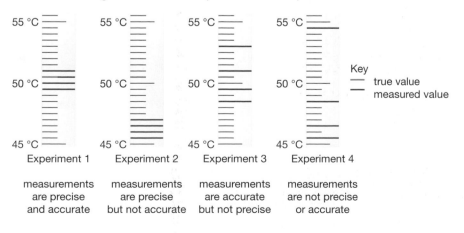

Experiment 1	Experiment 2	Experiment 3	Experiment 4
measurements are precise and accurate	measurements are precise but not accurate	measurements are accurate but not precise	measurements are not precise or accurate

Key
— true value
— measured value

235

Maths skills

The aim of science is to develop good explanations for observations of the natural world. Scientific explanations are based on data. Making sense of the data requires mathematical skills, including making calculations, and presenting and reading graphs and charts. This section of the book will support you in making sense of your own data from experiments and also the data presented by others.

Throughout the book are worked examples to remind you how to apply your mathematical skills in science contexts.

Numbers and units

At the heart of most scientific enquiries are measurements. Measurements are stated as a number and unit. When doing calculations it is important to be consistent in the units you use.

In science we use the SI system of measurements: the base units of kilogram (kg), metre (m), second (s), ampere (A), kelvin (K), and mole (mol) together with derived units including metre per second (m/s), newton (N), metres cubed (m^3), and degrees Celsius (°C).

These are the units you will use in your biology studies:

Quantity	Unit
mass	kilogram (kg)
length	metre (m)
time	second (s)
area	metre squared (m^2)
volume	metre cubed (m^3), decimetre cubed (dm^3)
energy	joule (J), kilowatt-hour (kWh)
temperature	degrees Celsius (°C), kelvin (K)
body mass index (BMI)	kilogram per metre squared (kg/m^2)
light intensity	watt per metre squared (W/m^2)

H

Standard form

Sometimes the numbers used in scientific measurements and calculations are very large or very small. For example, the distance from the Earth to the Sun is 146 900 000 000 m.

This is difficult to read and it can be easy to 'lose' one of the zeroes in a calculation. It is more convenient to express the number in standard form.

A number written in standard form has two parts:

the multiplier (10 raised to the power needed to give the correct value of the number)

a decimal number more than one, less than ten

1.469×10^{11}

Worked examples: Converting large and small numbers to standard form

1 *It is estimated that there are 8 700 000 species on Earth. Write this number in standard form.*

Step 1: Find the decimal number that is more than one and less than 10.

8.7

Step 2: Find how many times you need to *multiply* 8.7 by 10 to get 8 700 000

$8\ 700\ 000 = 8.7 \times 10 \times 10 \times 10 \times 10 \times 10 \times 10$

> Multiply by 10 six times, which is the same as multiplying by 10^6.

Answer:

$8\ 700\ 000 = 8.7 \times 10^6$

2 *The diameter of a white blood cell is about 0.000012 m. Write this quantity in standard form.*

Step 1: Find the decimal number that is more than one and less than 10.

1.2

Step 2: Find how many times you need to *divide* 1.2 by 10 to get 0.000012.

$$0.000012 = 1.2 \times \frac{1}{10} \times \frac{1}{10} \times \frac{1}{10} \times \frac{1}{10} \times \frac{1}{10}$$
$$= 1.2 \times \frac{1}{10^5}$$
$$= 1.2 \times 10^{-5}$$

> Divide by 10 five times, which is the same as multiplying by 10^{-5}.

Remember, dividing by 10 is the same as multiplying by $\frac{1}{10}$

> The negative power shows the number is *less* than 1.

Answer:

$0.000012\ m = 1.2 \times 10^{-5}\ m$

You should make sure you know how to work with numbers in standard form on your calculator. Different calculators have different labels on buttons for '10 raised to the power of', for example:

| EE | or | EXP | or | × 10ˣ |

Prefixes for units

Sometimes prefixes for units are used instead of writing the quantity in standard form.

For example, the diameter of a white blood cell is 1.2×10^{-5} m, which is 12 µm.

The table shows the prefixes that are used with scientific quantities.

nano	micro	milli	centi	deci	kilo	mega	giga	tera
n	µ	m	c	d	k	M	G	T
0.000 000 001	0.000 001	0.001	0.01	0.1	1000	1 000 000	1 000 000 000	1 000 000 000 000
$\times 10^{-9}$	$\times 10^{-6}$	$\times 10^{-3}$	$\times 10^{-2}$	$\times 10^{-1}$	$\times 10^{3}$	$\times 10^{6}$	$\times 10^{9}$	$\times 10^{12}$

Order of magnitude

Rounding a number to the nearest **order of magnitude** means rounding the number to the nearest power of 10.

Worked example: Order of magnitude

The radius of a carbon atom is measured to be about 0.07 nm.
What is this in metres to the nearest order of magnitude?

Step 1: Write down the length in metres.

$0.07 \text{ nm} = 0.07 \times 10^{-9} \text{ m} = 7 \times 10^{-11} \text{ m}$

> 7 is closer to 10 than to 1

Step 2: Write down the value to the nearest power of 10

$7 \times 10^{-11} \text{ m} \approx 10 \times 10^{-11} \text{ m}$

$10 \times 10^{-11} \text{ m} = 1 \times 10^{-10} = 10^{-10}$

Answer:

$7 \times 10^{-11} \text{ m} \approx 10^{-10} \text{ m}$ (to nearest order of magnitude)

Sometimes when two measurements are compared we simply want to know if they are in the same order of magnitude. If two numbers differ by an order of magnitude, then one number is about 10 times bigger than the other.

Worked example: Comparing orders of magnitude

The radius of a hydrogen atom is measured to be about 0.025 nm. The radius of a lead atom is measured to be about 0.18 nm. Compare the size of hydrogen atoms and lead atoms, and decide if their sizes are the same order of magnitude.

Step 1: Write down what you know, ensuring both measurements are in the same units.

hydrogen: 0.025 nm
lead: 0.18 nm

Step 2: Divide the larger number by the smaller number.

$\dfrac{\text{larger value}}{\text{smaller value}} = \dfrac{0.18 \text{ nm}}{0.025 \text{ nm}} = 7.2$

Answer:

Lead atom is 7.2 times larger, which is less than 10 times larger – this means they are the same order of magnitude.

Significant figures

Data from measurements in scientific experiments should show the **precision** of the measurement. The number of **significant figures** shows the precision that can be claimed for a piece of data.

The first significant figure in a number tells you the approximate size of the number. The first significant figure in a number is the first non-zero digit from the left.

For example, in the number 5437 the first significant figure is the 5; this tells you that the value is more than 5000 and less than 6000. You can round the number to any number of significant figures. So this number is 5000 (to 1 significant figure), 5400 (to 2 significant figures), or 5440 (to 3 significant figures).

The same principle applies to numbers less than one. In the number 0.0342 the first significant figure is 3, so we know the number is between three-hundredths and four-hundredths. The value is 0.03 (to 1 significant figure) or 0.034 (to 2 significant figures).

You may see significant figures abbreviated to 'sig. fig.' or 'S.F.'.

Significant figures are a more useful way of expressing precision than decimal places. For example, a length measured with a metre ruler is made to the nearest millimetre (thousandth of a metre). The length of an A4 sheet of paper can be written as 297 mm, 29.7 cm, 0.297 m, and 2.97×10^{-1} m. All of those measurements are given to the same number of significant figures, even if the number of decimal places is different.

A calculated value should be given to the same number of significant figures as the least precise measurement in the data.

Worked example: Significant figures

Calculate the area of a rectangle with sides 26 mm and 13 mm. Give your answer to 2 significant figures.

Step 1: Write down what you know.

length = 26 mm
breadth = 13 mm

Step 2: Write down the equation you will use.

area of rectangle = length x breadth

Step 3: Substitute in the numbers and calculate the area.

area = 26 mm x 13 mm
= 338 mm²

Step 4: Start at the first non-zero digit and count 2 significant figures.

338 mm²

Step 5: Apply the 5-or-more rounding rule to the next digit.

area = 340 mm²

Answer:

area = 340 mm² (2 significant figures)

Making sense of data

Once data has been collected from an experiment it can be processed to show any patterns. This processing often means doing calculations and plotting graphs.

Using percentages

Percentages appear everywhere, for example, in interest rates on loans, discounts in the sales, and, of course, in science. Percentage is a way of describing a fraction of something. 'Per cent' (symbol %) means 'per hundred'. If 20 people out of a group of 100 people are left-handed, then we could say that 20% are left handed – and therefore that 80% are right-handed or ambidextrous.

If one quarter of a population have blonde hair we could say that 25 out of every 100 people have blonde hair – that is, 25% have blonde hair.

Percentages can be used to make comparisons between sets of data, to calculate efficiency, or to calculate the yield of a process.

Worked example: Calculating percentages

A student carries out an experiment to find the best way to germinate seeds. One seed tray has 80 seeds and 60 of them germinate to produce plants. A second tray has 90 seeds and 70 of them germinate. Which tray gives the better yield of plants?

Step 1: Write down what you know.

Tray 1:	Tray 2:
number of seeds = 80	number of seeds = 90
number of plants = 60	number of plants = 70

Step 2: Write down an equation to work out the percentage yield.

$$\text{yield} = \frac{\text{number of plants}}{\text{number of seeds}} \times 100\%$$

Step 3: Write down an equation to work out the percentage yield.

$$\text{yield 1} = \frac{60}{80} \times 100\% \qquad \text{yield 2} = \frac{70}{90} \times 100\%$$
$$= 0.75 \times 100\% \qquad\qquad = 0.78 \times 100\%$$
$$= 75\% \qquad\qquad\qquad = 78\%$$

Answer: Tray 2 had a slightly better yield (78%) than tray 1 (75%).

> Tray 2 produced more plants, but there were more seeds so percentage yields make it easier to compare.

Finding the best estimate

The **range** of a data set describes the spread of data. For example, in a class of students the heights of the students may have a range from 155 cm to 185 cm.

The **average** of a data set is the single number that best represents the data set. There are different ways of representing that average.

The most commonly used average is the **mean value**. The mean value is found by adding up a set of measurements and then dividing by the number of measurements.

$$\text{mean} = \frac{\text{sum of all the measurements}}{\text{number of measurements}}$$

Worked example: Finding the mean

A group of students measure their heights. The measurements are: 160 cm, 165 cm, 167 cm, 168 cm, and 185 cm. Calculate the mean height.

Step 1: Write down what you know.

heights: 160 cm, 165 cm, 167 cm, 168 cm, and 185 cm
number of students: 5

Step 2: Write down an equation to work out the mean height.

$$\text{mean} = \frac{\text{sum of all the measurements}}{\text{number of measurements}}$$

Step 3: Substitute in the values to calculate the percentage yield.

$$\text{mean} = \frac{160 \text{ cm} + 165 \text{ cm} + 167 \text{ cm} + 168 \text{ cm} + 185 \text{ cm}}{5}$$
$$= \frac{845 \text{ cm}}{5}$$
$$= 169 \text{ cm}$$

Answer: mean height = 169 cm

Sometimes the mean is not a good representation of the data. In the example of the group of students above, only one of the students is taller than the mean value. The 185 cm measurement has distorted the calculation because it is so much larger than the others.

A better representative value is sometimes the **median**: the middle value. In this data set, the median is167 cm.

In a large data set with many values, the **mode** might be the best representative value – that is the value that occurs most often.

Whether you choose the mean, the median, or the mode, it is always good to also give the range over which the data was spread.

Making sense of graphs

Scientists use graphs and charts to present data clearly and to look for patterns in the data. You will need to plot graphs or draw charts to present data, and then describe the patterns or trends and suggest explanations for them. Examination questions may also give you a graph and ask you to describe and explain it.

Reading the axes

Look at the two charts in the margin, which both provide data about daily energy use in several countries.

On the first chart the value for China is greater than for the US. But on the second chart the value for the US is much greater than for China. Why are the charts so different if they both represent information about energy use?

Look at the labels on the axes.

One shows the *energy use per person per day*, the other shows the *energy use per day by the whole country.*

The first graph shows that China uses a similar amount of energy to the US. But the population of China is much greater so the energy use per person is much less.

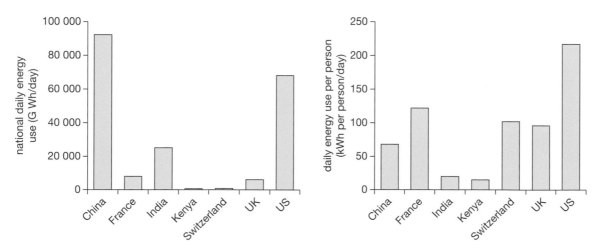

Graphs to show daily energy use in a range of countries, total and per person.

First rule of reading graphs: Read the axes and check the units.

Describing the relationship between variables

The pattern of points plotted on a graph shows whether two factors are related.

Look at the graph of how the mass of a baby changes in the first 12 weeks after birth.

The two variables are age and mass. The graph tells a story about the relationship between these two variables – the baby gets a little lighter in the first two weeks and then heavier. But we can describe the pattern in more detail than that. Between three weeks and nine weeks her mass increases steadily, then increases less quickly up to 12 weeks. The slope of the graph – the **gradient** – is constant between three weeks and nine weeks.

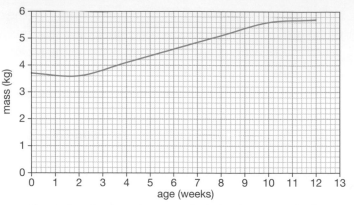

Graph showing how the mass of a baby changes in the first 12 weeks after birth.

Calculating the gradient of a graph

Many graphs show how a quantity changes with time – it might be the size of a population, the height of a plant, the concentration of a substance, or the speed of a moving object. Time is plotted on the **x**-axis and the changing quantity being measured is plotted on the *y*-axis. The gradient of such graphs describes the **rate** of change of the quantity with time.

$$\text{rate of change of quantity} = \frac{\text{change in quantity (}y\text{-axis)}}{\text{time taken to change (}x\text{-axis)}}$$

Worked example: Finding the gradient of a straight-line graph

The graph shows how a baby's mass changed in the weeks after she was born. Use the graph to work out the average rate at which her mass increased between three weeks and nine weeks.

Step 1: Use crosses (X) to mark on the line the points the question is asking about.

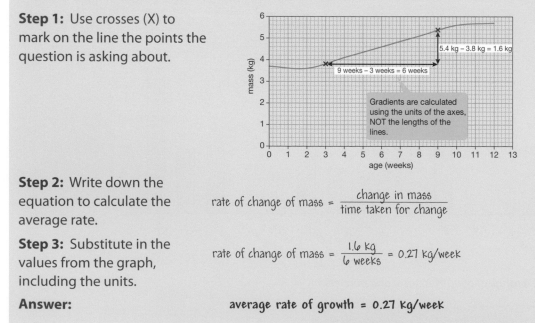

5.4 kg – 3.8 kg = 1.6 kg

9 weeks – 3 weeks = 6 weeks

Gradients are calculated using the units of the axes, NOT the lengths of the lines.

Step 2: Write down the equation to calculate the average rate.

$$\text{rate of change of mass} = \frac{\text{change in mass}}{\text{time taken for change}}$$

Step 3: Substitute in the values from the graph, including the units.

$$\text{rate of change of mass} = \frac{1.6 \text{ kg}}{6 \text{ weeks}} = 0.27 \text{ kg/week}$$

Answer: average rate of growth = 0.27 kg/week

Second rule of reading graphs: Describe each phase of the graph, including ideas about the meaning of the gradient, and other data including units.

Is there a correlation?

Sometimes we are interested in whether one thing changes when another does. If a change in one factor goes together with a change in something else, we say that the two things are correlated.

The two graphs on the right show how global temperatures have changed over time and how levels of carbon dioxide in the atmosphere have changed over time.

Is there a **correlation** between the two sets of data?

Look at the graphs – why is it difficult to decide if there is a correlation?

The two sets of data are over different periods of time, so although both graphs show a rise with time, it is difficult to see if there is a correlation.

It would be easier to identify a correlation if both sets of data were plotted for the same time period and placed one above the other, or on the same axes, like this:

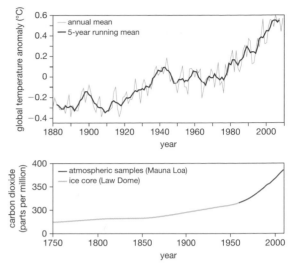

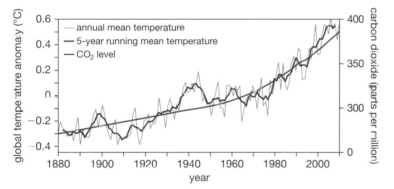

Graphs to show increasing global temperatures and carbon dioxide levels. Source: NASA.

When there are two sets of data on the same axes, take care to look at which axis relates to which line.

Another way to check for a correlation is to plot the two variables on a scatter graph. Look at the scatter graph in the margin. Is there evidence that the two variables:

- are correlated?
- show a causal relationship?

Explaining graphs

Explaining the patterns or trends shown by a graph is different to describing them. It requires us to use scientific ideas to suggest what could be causing the observed patterns.

When a graph suggests that there is a correlation between two sets of data, scientists try to find out if a change in one factor *causes* a change in the other. They look for a mechanism that explains how one factor affects the other.

Displaying frequency data

Frequency data shows the number of times a value occurs. For example, if four students have a pulse rate of 86 beats per minute, then the data value 86 has a frequency of four.

Frequency data is presented using a bar chart or a histogram. It's important to know the difference between these two types of chart and when to use each one.

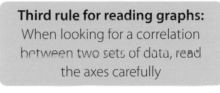

> **Third rule for reading graphs:**
> When looking for a correlation between two sets of data, read the axes carefully

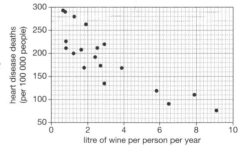

Graph showing that heart disease is less common in people who drink a moderate amount of wine, from a study in over 19 countries.

Bar chart: discontinuous data

Country	Number of nuclear power plants
USA	99
France	58
Japan	43
Russia	34
China	27
UK	16

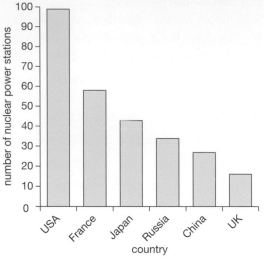

The data in the table has been plotted as a bar chart, showing the number of working nuclear power stations in different countries in June 2015.

Look at the bar chart displaying the number of nuclear power stations per country. Each country is separate, and it's not possible to measure a value part way between any two countries. The data are said to be discontinuous. A bar chart is used to display discontinuous frequency data.

A bar chart would also be used to summarise the number of trees of each species in a wood (species names on the x-axis), and the number of students studying chemistry in different schools (school names on the x-axis).

In a bar chart all of the bars should be drawn with equal width, and there should be a gap between each bar.

Histogram: continuous data

You may work with frequency data for a continuous variable, for example, height. Continuous data can be divided into groups known as class intervals. Collecting data in class intervals can be done by tallying. As a rule of thumb, try to divide the data into at least five class intervals.

Look at the data recording the heights of 31 people:

Height (m)	Tally	Frequency
1.60–1.65	I	1
1.65–1.70	IIII	4
1.70–1.75	IIII IIII II	12
1.75–1.80	IIII III	8
1.80–1.85	IIII	5
1.85–1.90	I	1

Note that the class interval 1.60–1.65 includes all the people with a height that is:

- greater than or equal to 1.60 m
- less than 1.65 m.

This means that a person who is 1.64 m tall is included in the class interval 1.60–1.65, but a person who is 1.65 m tall is included in the next class interval (1.65–1.70).

A histogram is used to display continuous frequency data.

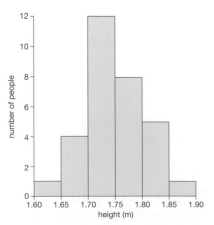

A histogram displaying the frequency of height in a group of 31 people.

Probability

It is often useful to make a quantitative prediction about how likely it is that a particular outcome will happen. For example, we may want to predict how probable it is that a child of two parents will inherit a particular feature.

The **probability** of a particular outcome occurring can be expressed as a decimal number between 0 and 1. If the probability is 0, the outcome is impossible. If the probability is 1, the outcome is certain.

The probability of a particular outcome can be calculated using the equation:

$$\text{probability} = \frac{\text{number of ways the particular outcome can happen}}{\text{total number of possible outcomes}}$$

Worked example: Calculating the probability of an outcome

One card will be picked at random from a standard pack of playing cards. What is the probability that the card will be a queen?

Step 1: Write down what you know.

number of cards in a pack = 52
number of queens in a pack = 4
particular outcome = card is a queen
number of ways the particular outcome can happen = 4
total number of possible outcomes = 52

Step 2: Write down the equation for the probability of an outcome.

$$\text{probability} = \frac{\text{number of ways the particular outcome can happen}}{\text{total number of possible outcomes}}$$

Step 3: Substitute what you know into the equation, and calculate the probability.

$$= \frac{4}{52}$$

Answer:

probability = 0.08 (to 1 significant figure)

There are four queens in a standard deck of 52 playing cards. We can calculate the probability of picking a queen when one card is picked at random.

Sampling

Sometimes we want to collect data about a large number of individuals or a large area. For example, we may want data about the body mass of men in England, or we may wish to know how many daisies are growing on a playing field. But it would take far too long to measure the mass of every man in England or to count every daisy on a playing field.

To overcome this problem, we can collect data about a **sample**. A sample is a proportion of the whole population or area we wish to study. For example, we may measure the masses of 1000 men from England, or count the number of daisies in a few square metres of the playing field.

Conclusions about a sample can only be applied to the whole population if it is a **representative sample**. The characteristics of a representative sample are very similar to the characteristics of the whole population. Choosing the sample at random, and using as large a sample as possible, can help to ensure it is representative.

It would take a long time to count all the plant species in an entire playing field. Instead, we can count the species in a small sample area.

Models in science

Find out about

- what a model is
- how models are used to investigate, explain, and predict
- the benefits and limitations of models
- representational, descriptive, and mathematical models

Key words

➤ model
➤ system
➤ representational model

Most children like to play with toys. A popular toy is a model car. It's called a **model** because it represents something in the real world. It has some of the main features of a real car, such as wheels, doors, a windscreen, and a roof. But some features, such as the engine, are not included. The model shows that the car can move when the wheels turn.

A model car is a simple way to represent the main features of a car, but much of the detail of a real car is not included in the model.

Models are everywhere

People use models all the time, even if they don't realise it. A model isn't always an object, such as a toy car. Models can be words, pictures, and numbers. For example, a map is a model.

Millions of people visit London every year, and they use maps to help them travel around. The usual map of the London Underground is a useful guide for getting from one Tube station to another. It's a good model of the city's underground train network because it shows how the stations and lines are connected. It is quick and easy to understand. It can answer questions such as: 'What is the most direct way to get from Bond Street to Westminster?'

We can also use the model to make predictions. For example, we could predict that it would take longer to travel by Tube to Westminster from Camden Road than it would from Bond Street.

However, the London Underground map can't solve all of a traveller's problems. It doesn't show how the Tube stations relate to streets and landmarks on the surface. These features are not included in the model. To answer the question 'What is there to see around Westminster tube station?' we can use a different, more detailed model – for example, a street map.

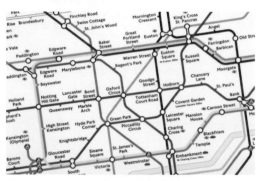

This map is a simple model of the London Underground train network. It helps people to work out how to get from one Tube station to another.

This map is another model of London, but it includes features such as streets and local landmarks.

What is a model?

From the everyday examples of model cars and maps, we know that a model:

- is a simpler way of representing something in the real world
- includes some, but not necessarily all, of the features of the thing it represents
- can show how these features are connected or interact
- can be used to explain things, answer questions, and make predictions.

We also know that the usefulness of a model is limited by how closely it represents the real world and that different models are useful in different situations.

Usually, a model represents a collection of interacting parts. Scientists often refer to a collection of interacting parts as a **system**.

Using models in science

Models are useful in science. They help us to explain how things work and interact. They also help us to make predictions and investigate possible outcomes. Scientific models can be words, pictures, objects, numbers, graphs, or equations.

Many different models are used in science, but they can be grouped into three main types. These are described below. You've already used models in your science lessons. An example of a model that you should be familiar with is given for each type.

A number of scientific models are highlighted throughout this book – look out for the red boxes like the ones shown here.

Representational models

A **representational model** uses simple shapes or analogies to represent the interacting parts of a system. One example of a representational model is the particle model – it helps us to visualise the tiny particles (atoms and molecules) that make up substances.

Different models of the same thing can contain different amounts of detail. A model car made of blocks will roll on its wheels from the top of a slope to the bottom. A remote-controlled car contains a motor, so can be driven from the bottom of a slope to the top. The models can be used to demonstrate and investigate different behaviours.

The particle model of matter

All matter is made of very tiny particles with attractive forces between them. The particle model represents the particles as spheres. It helps us explain the arrangement and movement of the particles in different states of matter.

However, this model makes some simplifications. For example, particles of matter are not all identical spheres – they have different shapes, sizes, and masses. This means there are limitations to the model and what we can do with it.

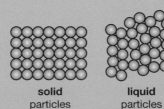

solid
particles
touching and

liquid
particles
touching but

gas
particles far
apart and can

You should already know about some simple models of photosynthesis (the process plant cells use to make food).

A speed camera takes photographs of a car as it travels over markings on the road. The time between the photographs and the distance between the markings are used to calculate the car's speed.

Descriptive models

A **descriptive model** uses words to identify features of a system and describe how they interact. One example of a descriptive model is a simple account of the inputs and outputs of photosynthesis. It helps us to explain how plant cells can make their own food in the form of glucose (a type of sugar).

A simple model of photosynthesis

Photosynthesis is a complicated biological process, involving a number of chemical reactions inside cells. However, a simple description of the inputs and outputs can help us to understand what is going on.

Inputs	Outputs
light	glucose
water	oxygen
carbon dioxide	

This is a very simple way of summing up the process of photosynthesis. We could use this simple model to make a prediction, for example, that increasing the supply of inputs would increase the amount of glucose made.

However, this model misses out much of the detail. It does not tell us anything about the chemical reactions or where they take place.

Mathematical models

A **mathematical model** uses patterns in data of past events, along with known scientific relationships, to predict what will happen to one variable when another is changed. This involves doing calculations. If the calculations are very complicated, they can be done by a computer – this is a **computational model**. But a very simple example of a mathematical model is the relationship between speed, distance, and time.

Modelling the speed of a moving object

There is a simple, scientific relationship between the speed of a moving object, the distance it moves, and the time taken to move that distance. The relationship is:

$$\text{speed} = \frac{\text{distance}}{\text{time}}$$

This is a mathematical model. We can use it to calculate the speed of a moving object if we have values for the other two variables (distance and time). We can also use the model to make a prediction, for example, if the car travels for a longer time at the same speed, it will travel further.

But this simple model is limited. It can only be used to calculate an average speed from the total distance and time – it does not include information about changes in speed, such as acceleration or deceleration. Also, it does not include any information about the direc tion the car is travelling.

One type of mathematical model is a **spatial model**. A computer is used to make a model of one or more objects in a three-dimensional space. The model can be used to predict the outcome of changing a variable (e.g., temperature) in a given space (e.g., a landscape and the atmosphere above it). The space can be divided into sections and outcomes predicted for each section. Spatial models are often used to make predictions about weather and climate.

Why use models?

If we tried to describe and explain every unique situation in the world, we would never get finished! This is one reason why models are useful. A model is a general explanation that applies to a large number of situations – perhaps to all possible ones. For example, every cell in your body is different, but we can use a model of an animal cell to describe the main features that animal cells have in common.

A benefit of models is that they enable us to investigate situations that we cannot investigate using practical experiments – perhaps because it is not ethical, is too expensive, or is not possible to do so. For example, mathematical models enable us to investigate the future effects of human activities on the Earth's climate and biodiversity. This helps us to make predictions about the likely outcomes of different courses of action, which can affect the decisions we make.

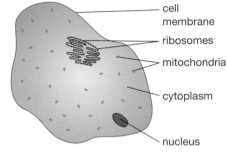

A simple model of an animal cell. Every animal cell is different, but the model includes the main features they have in common.

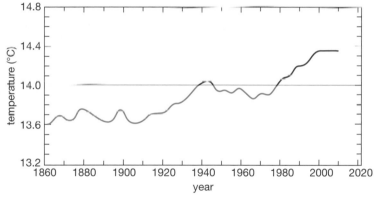

Measurements taken over the past 150 years show that the Earth's temperature is rising. We can use a mathematical model to predict how it may change in the future.

Models are very useful but we must always remember that they are limited. Even a very good model cannot represent the real world exactly, so outcomes in the real world could be different from a model's outcomes. Models help weather forecasters make predictions about the weather, but they don't always get it exactly right. Always think carefully about how much confidence you can have in claims based on a model – be realistic, and don't be too surprised if things turn out a little differently in the real world.

Key words

➤ descriptive model
➤ mathematical model
➤ computational model
➤ spatial model

Appendices

Mathematical relationships

You will need to be able to use these mathematical relationships.

Body mass index (BMI):

$$\text{body mass index (kg/m}^2) = \frac{\text{body mass (kg)}}{[\text{height (m)}]^2}$$

Area A of circle with radius r:

$$A = \pi r^2$$

Estimated population size from capture-mark-recapture samples:

$$\text{estimated population size} = \frac{\text{number of individuals given marks} \times \text{number of individuals recaptured}}{\text{number of recaptured individuals that have a mark}}$$

Rate of chemical reaction:

$$\text{mean rate of reaction (cm}^3\text{/s)} = \frac{\text{change in volume of gas (cm}^3)}{\text{time for change to happen (s)}}$$

$$\text{mean rate of reaction (g/s)} = \frac{\text{change in mass (g)}}{\text{time for change to happen (s)}}$$

Relative light intensity LI at distance d from a point light source:

$$LI \propto \frac{1}{d^2}$$

Rate of water uptake by a plant in a potometer:

$$\text{rate of water uptake (cm}^3\text{/s)} = \frac{\text{change in volume of water (cm}^3)}{\text{time taken for volume to change (s)}}$$

Useful chemical equations

Photosynthesis

$$6\,CO_2(g) + 6H_2O(l) \longrightarrow C_6H_{12}O_6(aq) + 6O_2(g)$$

Aerobic respiration

$$C_6H_{12}O_6(aq) + 6O_2(g) \longrightarrow 6CO_2(g) + 6H_2O(l)$$

Glossary

abiotic The non-living parts of an ecosystem (e.g., the atmosphere, water, soil, and rocks).

abiotic factor A non-living factor that can affect the distribution and abundance of living organisms (e.g., temperature, light intensity, soil pH, and the availability of water).

abundance How many organisms are found in each part of an ecosystem.

accuracy How close a quantitative result is to the true or 'actual' value.

active site The part of an enzyme where the chemical reaction takes place. The reacting molecules (substrates) fit into the active site.

active transport Molecules are moved in or out of a cell using energy. This process is used when transport needs to be faster than diffusion, and when molecules are being moved from a region where they are at low concentration to where they are at high concentration.

adaptation A feature that helps an organism survive in its environment.

ADH Antidiuretic hormone, which is secreted by the pituitary gland. It controls the reabsorption of water in the kidneys to control the water balance of the body.

adrenal gland Gland that secretes the hormones adrenaline and cortisol.

adrenaline Hormone secreted by the adrenal gland at times of stress. It prepares the body for action.

adult stem cell A stem cell that can only differentiate to make one or a small number of cell types.

advantage An individual organism has an advantage if it has a feature that enables it to survive and reproduce more successfully than other individuals.

aerobic respiration A type of cellular respiration that uses oxygen.

AIDS Acquired immune deficiency syndrome, an STI caused by the HIV virus. The body's immune system is attacked by the virus and gradually becomes weakened.

allele The two copies of a gene in a pair of chromosomes are called alleles. They can be the same or different.

alveolus (plural alveoli) Human lungs are divided up into millions of tiny sacs called alveoli. They form an exchange surface with a very large surface area for gaseous exchange.

amino acid Amino acids are joined in long chains to make proteins. All the proteins in living organisms are made from 20 different amino acids joined in different orders.

anaerobic respiration A type of cellular respiration that does not use oxygen.

antagonistic Actions or effectors that have opposite effects.

antibiotic resistance The ability of bacteria to survive exposure to antibiotics. The ability is caused by mutations in the bacteria's genes.

antibiotics Substances that kill or stop the growth of bacteria, but which do not work against fungi or viruses.

antibodies Protein molecules made by white blood cells to fight pathogens; each antibody only recognises one specific antigen.

antigens Molecules on the surface of cells and pathogens.

archaea One of the domains used in modern classification systems, containing prokaryotic organisms.

artery Type of blood vessel that carries blood from the heart to cells and tissues.

ATP Adenosine triphosphate, a substance made by cellular respiration. Cells need a constant supply of ATP for life processes. ATP is a chemical store of energy.

atrium (plural atria) The human heart is made of four muscular chambers. The atria are the upper chambers. They receive blood and pump it into the lower chambers (ventricles).

autotroph An organism that can make its own food.

average The average of a set of data is the single number that best represents the data. Commonly used averages include the mean, median, and mode.

axon A long, thin extension of the cytoplasm of a neuron. The axon carries nerve impulses very quickly.

bacteria (singular bacterium) Large group of single-celled, prokaryotic microorganisms. Some bacteria cause disease.

biconcave The shape of a red blood cell, like a disc that has been squeezed from both sides in the middle.

bioaccumulation An increase in the concentration of a substance in organisms in the higher trophic levels of a food chain.

biodegradable Substances that can be broken down by microorganisms, such as bacteria and fungi. Most paper and wood items are biodegradable, but most synthetic polymers, such as plastics, are not.

biodiversity The biodiversity of the Earth, or of a particular area, is the combination of the diversity of living organisms, the diversity of genetic material the organisms have, and the diversity of ecosystems in which the organisms live.

bioinformatics Using computers to process and analyse biological data, including genome sequences.

biological control The introduction of new species into an ecosystem to kill pests that carry plant pathogens and damage plants.

biomass All the tissues that make up an organism. Some of an organism's biomass is transferred to another organism when it is eaten.

biotic The living parts of an ecosystem, i.e., all the living organisms.

biotic factor A living factor that can affect the distribution and abundance of living organisms (e.g., a predator, a competitor, or pathogens).

bladder Organ that collects urine from the kidneys for excretion by urination.

blind trial A clinical trial in which the patient does not know whether they are taking the new medicine, but their doctor does.

body mass index (BMI) A calculated value that indicates whether a person's body mass is healthy for a person of their height.

capillary Type of tiny blood vessel that carries blood through tissues (from arteries to veins).

capture-mark-recapture A sampling technique used to estimate the size of a population of animals. A sample of individual animals is collected from the population, and each animal given an identifying mark. Later, another sample is collected. A mathematical equation is used to estimate the population size based on the number of marked animals in the second sample.

carbohydrate A substance made of carbon, hydrogen, and oxygen. Carbohydrates include sugars (e.g., glucose) and natural polymers (e.g., starch and cellulose).

carbon cycle The natural processes that recycle carbon through the living and non-living parts of an ecosystem.

carcinogen A substance that can cause damage to DNA. This can result in a mutation and can cause a cell to become a cancer cell.

cardiovascular diseases A group of non-communicable diseases affecting the blood vessels and heart.

carrier An individual who has a recessive variant but does not show the feature associated with the variant because they also have a dominant variant (they are heterozygous). They can pass the recessive variant to their offspring.

catalyst A catalyst is a substance that increases the rate of a chemical reaction, but which is left unchanged by the reaction.

cause When a change in a factor produces a particular outcome, and there is a mechanism to explain this link, then the factor is said to cause the outcome.

cell cycle Most cells go through a cycle of phases, including interphase and either one cell division (mitosis) or two cell divisions (meiosis).

cellular respiration A series of chemical reactions that happens in all living cells. Glucose is used up and ATP is made for the cell to use.

central nervous system (CNS) The brain and spinal cord.

chemical control The use of substances, usually sprayed onto crops, to reduce the spread of plant pathogens and pests that carry them.

chlorophyll A pigment that absorbs light and splits water into hydrogen and oxygen in the first stage of photosynthesis.

chloroplast An organelle containing chlorophyll, found in some plant cells. The reactions of photosynthesis happen in chloroplasts.

chromosome Long, thin, threadlike structure made from molecules of DNA. Chromosomes store genetic informaton.

circulatory system An organ system that includes the heart, blood vessels, and blood. The circulatory system transports substances around the body.

classification Placing organisms into groups based on their similarities and differences.

clinical testing A stage of drug development. A new medicine is tested on humans to find out whether it is safe and whether it works.

communicable disease A disease caused by an infection with a pathogen, which can be passed from one organism to another in body fluids, contamined food and water, or by direct contact.

community Interacting populations of organisms living in the same ecosystem.

competition Organisms that require the same resource (e.g., food, water, space, light, shelter, mates, pollinators, and seed dispersers) must compete for the resource.

complication A second disease that results from having a first disease.

computational model A type of mathematical model in which the calculations are done by a computer.

conservation (biodiversity) The protection of ecosystems and species to prevent (or slow the rate of) biodiversity loss.

consumer An organism that eats others in a food chain. A consumer is a heterotroph, because it cannot make its own food.

contamination The presence of an unwanted (and possibly harmful) substance or organism in the body, in a substance, or on a surface.

contraception Physical barrier or treatment that prevents pregnancy. Some forms of contraception, such as condoms, can also reduce the spread of STIs.

contraceptive pill Pill that contains hormones, such as progesterone, to disrupt the menstrual cycle and prevent pregnancy.

control group When testing a new treatment, a control group of people does not receive the new treatment. They may receive an existing treatment or a placebo.

computational model A type of mathematical model in which the calculations are done by a computer.

coronary arteries Blood vessels that supply blood carrying oxygen and glucose directly to the muscle cells of the heart.

coronary heart disease A type of cardiovascular disease in which fatty deposits cause blood clots in the coronary arteries, increasing the chance of a heart attack.

correlation When an outcome happens if a specific factor is present but does not happen when it is absent, or if a measured outcome increases (or decreases) as the value of a factor increases, there is a correlation between the two. For example, there is a correlation between pollen count and the number of hay fever cases.

crop rotation Crops are grown in different fields each year, so that each field is planted with a different crop each year.

decomposition Process in which dead organisms are broken down (decay) and substances in their bodies are returned to the environment.

deforestation Cutting down and clearing forests leaving bare ground. This damages ecosystems and causes biodiversity loss.

denatured When the shape of an enzyme has been changed, usually as a result of high temperature or a pH change. Denatured enzymes no longer work because the shape of the active site has changed.

deoxygenated blood Blood in which most or all of the haemoglobin in the red blood cells is not bound to oxygen.

depth of field When a microscope is focused on a structure being observed, objects outside the depth of field (above and below the structure) appear blurred.

descriptive model A type of scientific model that uses words to identify features of a system and to describe how they interact. One example of a descriptive model is a simple account of the inputs and outputs of photosynthesis.

development How an organism changes as it grows and matures. Its cells are organised into different tissues and organs, and they take on specific jobs.

differentiation The process by which a cell becomes specialised to take on a specific job.

diffusion Passive movement of molecules. The net movement is from a region of their higher concentration to a region of their lower concentration.

digestive system An organ system that includes the mouth, oesophagus, stomach, liver, pancreas, and intestines. The digestive system breaks down the biomass in the food we eat.

distribution Where particular organisms are found within an ecosystem.

DNA Deoxyribonucleic acid, a natural polymer made of nucleotides. DNA carries the genetic code, which controls how an organism develops and functions.

domain The largest group of organisms used in modern classification systems. The three domains are bacteria, archaea and eukaryotes (including animals, plants, fungi, and protists).

domesticated Organisms that have been selectively bred to live and work with humans.

dominant You only need to have one copy of a dominant genetic variant to have the feature it produces. A dominant genetic variant will always show its associated feature in the phenotype.

double helix The shape of the DNA molecule, with two strands twisted together.

double-blind trial A clinical trial in which neither the doctor nor the patient knows whether the patient is taking the new medicine.

ecosystem A community of organisms and the environment in which it lives (and with which it interacts).

ecosystem services Benefits that humans receive from ecosystems, including provisions (e.g., food, clean water, materials, and medicines), fertile soil and crop pollination, the control of climate, the breaking down of waste and cycling of substances, and cultural benefits (e.g., recreation, discovery, and aesthetics).

effector Muscle or gland that carries out the response to a stimulus.

electron A tiny, negatively charged particle, which is part of an atom. Electrons are found outside the atom's nucleus.

electron microscope A type of microscope that uses a beam of electrons, rather than a beam of light, to produce an image.

embryonic stem cells Unspecialised cells in an embryo before the eight-cell stage. They can divide and differentiate to make all the specialised cells in the body.

endocrine system A communication system in the body that uses hormones released from glands to control cells and tissues elsewhere in the body.

endothermic An endothermic process transfers energy from its surroundings, making them cooler.

energy stores An energy store is something (such as the Sun or a chemical substance) that enables you to account for the energy at the start and end of a transfer by doing a calculation.

enzyme A protein with an active site that catalyses (speeds up) a chemical reaction.

epidemic An outbreak of a disease affecting many individuals in a population.

ethical Concerned with what is right or wrong. Ethical questions cannot be answered by science alone.

eukaryotic organism A type of organism that has cells with a nucleus and organelles with membranes. Includes all animals, plants, fungi, and protists.

eutrophication A type of environmental damage. Excess nitrate or phosphate in water causes rapid growth of algae, followed by death and decomposition of water plants (due to lack of light), and a reduction of oxygen levels in the water. This usually leads to the death of water animals (due to lack of oxygen).

evolution The process by which species gradually change over time. Evolution can produce new species.

exchange surface Surface where substances are exchanged between an organism and its external environment.

excretory system An organ system that includes the lungs, kidneys, bladder, intestines, rectum, and anus. The excretory system gets rid of undigested food, excess water, urea (in urine), and carbon dioxide.

exothermic An exothermic process transfers energy to its surroundings, making them warmer.

extinction The permanent disappearance of a species because all the members of the species have died.

extrapolation The process of extending the line of a graph to estimate values beyond the original data.

factor A variable that changes and may affect something else (the outcome).

false negative A wrong test result. The test result says that a person does not have a medical condition but this is incorrect.

false positive A wrong test result. The test result says that a person has a medical condition but this is incorrect.

family tree A genetic diagram that is used as a model of inheritance. It shows which members of a family have inherited a particular feature.

fatty sheath Insulating layer of fat around the outside of an axon.

fermentation Anaerobic respiration in yeast cells, which breaks down glucose, and makes ethanol (alcohol) and carbon dioxide.

fertile soil Soil in which crops and other plants can be grown.

fertilisation The fusing of gametes during sexual reproduction.

food chain A model of the feeding relationships between populations of organisms. A food chain shows one route of biomass transfer through a community.

food web A model of the feeding relationships between populations of organisms. A food web shows multiple routes of biomass transfer through a community.

fossil The remains of an organism that lived millions of years ago that has been preserved in rock.

fossil record A collection of millions of fossils, which provides evidence of the evolution of species.

FSH Follicle-stimulating hormone, which is important in controlling the menstrual cycle.

fungi (singular fungus) Large group of eukaryotic organisms that cannot make their own food. Most are multicellular, but some (yeasts) are single-celled. Some fungi cause disease.

gamete Sex cell used in sexual reproduction. Male and female gametes fuse during fertilisation. In humans, the male gamete is the sperm and the female gamete is the ovum (egg cell).

gas exchange The exchange of oxygen and carbon dioxide between an organism and its surroundings. Occurs in the lungs of humans and the stomata of plants.

gaseous exchange system An organ system that includes the nose and mouth, trachea, and lungs. The gaseous exchange system takes in oxygen from the air we breathe in, and excretes carbon dioxide into the air we breathe out.

gender An individual's gender depends on whether they feel masculine or feminine, and whether they are happy to identify themselves as a man or a woman. It is affected by a person's feelings, society, and biological factors such as their genome and hormones.

gene A region of DNA containing the instructions for a cell that give the order of amino acids in a protein.

gene expression When the instructions in genes are read to make proteins. Controlling gene expression controls which proteins are made by a cell.

gene technology Techniques such as genetic testing and genetic engineering that are based on our understanding of the genome.

genetic engineering Altering the characteristics of an organism by modifying the DNA in its genome.

genetic testing Testing an individual for particular genetic variants, including those that cause diseases.

genetic variant A different version of a gene, caused by a change (mutation) in the DNA.

genome The entire genetic material of an organism.

genome sequencing A technique that works out the complete sequence of bases in an individual's DNA.

genomics The study of the structure and function of genomes, which includes genome sequencing and bioinformatics.

genotype The genetic variants that an organism has make up its genotype.

genus A group used in the classification of organisms, containing several similar species.

gland Part of the body that makes hormones, enzymes, and other secretions (e.g., the pituitary gland, the salivary gland, and sweat glands).

globalisation The increase of international travel, trade, and cooperation, including the transportation of people and cargo (including plants and animals) around the world.

glucagon Hormone secreted by the pancreas that causes the liver to break down carbohydrate to release glucose.

gradient (of a graph) The gradient, or slope, of a graph is a measure of its steepness. It is calculated by choosing two points on the graph and calculating:

$$\frac{\text{the change in the value of the } y\text{-axis variable}}{\text{the change in the value of the } x\text{-axis variable}}$$

graduation A line on a container, ruler, or meter that marks a measurement.

graft-versus-host disease When bone marrow stem cells have been donated to a patient, the new bone marrow makes white blood cells. The white blood cells from the new bone marrow recognise the patient's body cells as non-self, and attack them.

greenhouse gas A gas that contributes to the greenhouse effect, in which the atmosphere absorbs infrared radiation from the Earth's surface and radiates some of it back to the surface, making it warmer than it would otherwise be. Greenhouse gases include carbon dioxide, methane, and water vapour.

habitat The place where an organism lives within an ecosystem. The habitat provides conditions in which the organism can thrive.

haemoglobin Protein molecule in red blood cells. Haemoglobin binds to oxygen and carries it around the body. It also gives blood its red colour.

hazard A source of potential harm to health or the environment.

heart attack Death of heart muscle that occurs when a coronary artery is blocked, starving the heart muscle cells of oxygen and food.

herd immunity The protection given to a population against an outbreak of a disease when a very high percentage of the individuals in the population have been vaccinated against the disease.

heterotroph An organism that cannot make its own food, so must eat other organisms.

heterozygous The two alleles of a gene are different genetic variants.

HIV Human immunodeficiency virus, the virus that causes AIDS.

homeostasis The processes in your body that constantly work to keep the body's internal environment the same.

homozygous The two alleles of a gene are the same genetic variant.

hormone A substance secreted by specialised cells in animals and plants. Hormones bring about changes in cells or tissues in different parts of the animal or plant.

hygiene Behaviours and conditions that reduce the spread of pathogens.

hypothesis A tentative explanation for an observation. A hypothesis is used to make a prediction that can be tested.

identification key A guide that helps to identify the species of an organism. A key may include a series of steps or questions, for example, about flower colour or the arrangement of leaves. It may also include pictures.

immune system A system of tissues and cells in an organism that fights infections using non-specific defences and (in animals) white blood cells that respond to specific pathogens.

immunity The ability to produce antibodies against a pathogen very quickly, to destroy it before it causes symptoms of disease. Immunity depends on memory cells.

incubation period The delay between infection with a pathogen and the appearance of symptoms of disease.

indicator species A species whose abundance can be used to indicate the level of pollution in an area.

industrialisation The process of change in a society, from farming and hand-production methods to mass production using machines and chemical processes.

infertility A state in which an organism is not able to reproduce using sexual reproduction.

inherited A feature is inherited when genetic information that affects the feature is passed from parents to offspring in the DNA of gametes.

insulin Hormone secreted by the pancreas that helps to control the level of sugar (glucose) in the blood. It causes cells to absorb glucose from the blood.

interdependence The size of a population of organisms affects, and is affected by, the sizes of other populations in an ecosystem. Feeding relationships are one way in which populations are interdependent.

interphase The longest phase of the cell cycle, in which the cell grows larger, organelles are copied, and each chromosome is copied.

interpolation The process of taking a pair of values from a graph that are in between the data points that were plotted.

inverse square law A mathematical relationship between light intensity and the distance from a point light source. The relative light intensity at any distance from the light source is inversely proportional to the square of the distance from the light source.

ionising radiation Radiation with sufficient energy to remove electrons from atoms in its path. Ionising radiation, such as ultraviolet, X-rays, and gamma rays, can damage living cells.

kick sampling A sampling technique used to collect and count organisms from a river.

kingdom A group used in the classification of organisms, containing millions of species. There are kingdoms of plants, animals, fungi, protists, and bacteria.

lactic acid The product of anaerobic respiration in animal cells. Lactic acid is toxic at high concentration and must be removed from the body quickly.

large intestine Tube-like organ that receives digested food from the small intestine, where water is absorbed (leaving faeces).

LH Luteinising hormone, which is important in controlling the menstrual cycle.

light intensity The amount of light reaching a given area in a given time.

light microscope A type of microscope that uses a beam of light to produce an image.

limiting factor The factor that prevents the rate of a reaction or process from increasing. For photosynthesis, the limiting factor may be light intensity, temperature, carbon dioxide concentration, or water availability.

lipid A substance made from gycerol and fatty acids. Lipids are used for storage and making cell membranes.

liver Organ that breaks down protein, amino acids, and toxins from the blood. The liver also makes bile, which is released into the small intestine to aid digestion.

lock-and-key model In chemical reactions catalysed by enzymes, molecules taking part in the reaction fit exactly into the enzyme's active site. The active site will not fit other molecules – it is specific. This is like a key fitting into a lock.

lung Organ made of tissues that are adapted to absorb oxygen from the air and remove carbon dioxide from the blood.

magnification Use of a lens to make an object appear larger.

mathematical model A type of scientific model that uses patterns in data of past events, along with known scientific relationships, to predict what will happen to one variable when another is changed. A simple example of a mathematical model is the relationship between speed, distance, and time.

mean value A type of average, found by adding up a set of measurements and then dividing by the number of measurements. You can have more confidence in the mean of a set of measurements than in a single measurement. The mean can be used as the best estimate of the true value.

median A type of average. The median is the middle value in a set of values ordered from smallest to largest.

medicine A substance, or mixture of substances, that treats the symptoms and/or cause of a disease.

meiosis A type of cell division that halves the number of chromosomes to produce gametes.

memory cells White blood cells that stay in the blood after infection and create immunity. They respond very quickly by producing antibodies against an antigen the second time it is encountered.

meniscus The water surface in a narrow tube curves to form a meniscus.

menstrual cycle The cycle of changes in a woman's body associated with ovulation. The cycle takes approximately 28 days.

menstruation The shedding of part of the uterus lining each month as part of the menstrual cycle.

meristem A region of unspecialised cells in a plant, which can develop into any kind of specialised cell.

mitochondria Organelles in animal and plant cells where some of the reactions of cellular respiration take place.

mitosis A short phase of the cell cycle in which chromosome copies separate, the nucleus divides, and the cell divides in two.

mode A type of average. The mode is the value that occurs most often in a set of values.

model A scientific model is a way of representing something from the real world, such as a system of interacting parts. It includes some, but not necessarily all, of the features of the system it represents. It can show how these features are connected or interact, and can be used to explain scientific ideas, answer questions, and make predictions.

model organism An organism that is extensively studied by scientists to help develop explanations that may also apply to other species. Examples include *Drosophila* (fruit flies) and the tobacco mosaic virus.

motor neuron Neuron that carries nerve impulses from the CNS to an effector.

multicellular An organism made of two or (usually) more cells.

mutation A change in the DNA of an organism. It may or may not affect the organism's phenotype.

natural selection When individual organisms are better adapted to their environment they are more likely to survive and reproduce, passing on their features to the next generation.

negative feedback Actions that reverse any change in a system, so that the system returns to a steady state.

nerve impulse Electrical signal carried by a neuron.

net movement The overall movement of molecules by diffusion, when more molecules move in one direction than another. If equal numbers of molecules move in all directions, there is no net movement.

neuron A specialised cell in the nervous system that transmits nerve impulses.

non-communicable disease A disease not caused by an infection with a pathogen, but by a person's genes, their environment, or unhealthy lifestyle.

non-specific defences Physical, chemical, and bacterial defences that are part of the immune system; they are always present, rather than being produced in response to a specific pathogen.

nucleotide The monomer of DNA and RNA. Each nucleotide is made of a common sugar, a phosphate group, and a base.

nucleus (cell) Cell organelle that contains the chromosomes in cells of plants, animals, fungi, and protists.

obese A person who is very overweight is described as obese. An adult with a BMI greater than 30 kg/m^2 is classed as obese.

oesophagus Tube that joins the mouth to the stomach.

oestrogen Hormone that is important in puberty and in controlling the menstrual cycle.

open-label trial A clinical trial in which both the patient and their doctor know whether the patient is taking the new medicine.

optimum The most favourable condition for an enzyme, at which it catalyses a reaction at its fastest rate.

order of magnitude If two numbers differ by an order of magnitude, then one number is about ten times bigger than the other. A value given to the 'nearest order of magnitude' will be given to the nearest power of ten.

osmosis The diffusion of water molecules from an area where they are in high concentration (a dilute solution) to an area where they are at lower concentration (a concentrated solution) through a partially permeable membrane.

outcome A variable that changes as a result of something else changing.

outlier A measured result that seems very different from other repeat measurements, or from the value you would expect. The measurement should be treated as data, unless there is a reason to reject it (e.g., when it is known that a mistake was made when the measurement was taken or recorded).

ovulation The release of an ovum (egg cell) from a mature follicle in an ovary.

oxygenated blood Blood in which most or all of the haemoglobin in the red blood cells is bound to oxygen.

pancreas Organ that secretes digestive enzymes that are released into the small intestine.

pandemic An outbreak of a disease affecting many populations over a wide geographical area.

partially permeable membrane A membrane that acts as a barrier to some molecules but allows others to diffuse through freely.

passive A passive process does not require an input of energy. For example, diffusion is a passive process.

pathogen An organism that causes disease. Some bacteria, viruses, fungi, and protists are pathogens.

peer review The process whereby scientists who are experts in their field critically evaluate another scientist's scientific paper or idea before and after publication.

peripheral nervous system (PNS) The receptors, nerves, and effectors in the body that are not part of the CNS but are connected to it.

personalised medicine Treatment that is matched to a person's genome, to reduce the risk of harmful reactions and help the person stay healthy.

phenotype The features that result from the information in an organism's genome (and interaction with the environment) make up its phenotype.

phloem A plant tissue that transports sugar through a plant.

photosynthesis A series of chemical reactions in the cells of producers. Carbon dioxide and water are used to make glucose and oxygen. The process requires light.

pituitary gland Gland that secretes hormones to control other glands.

placebo A substance that looks exactly like a medicine, but has no medicine in it. A placebo is sometimes given to the control group in a clinical trial.

plasma (blood) The liquid part of blood.

plasmid Small ring of DNA found in bacteria. Plasmids are not part of a bacterium's main chromosome and contain extra genes.

platelets Components of the blood that help it clot at injury sites; they are fragments of cells made from the cytoplasm of large cells.

polyculture Growing a mixture of different plant species together.

population A group of organisms of the same species living in the same place.

pre-clinical testing A stage of drug development. A new medicine is tested on human cells grown in a laboratory and then on animals, to find out whether it is safe and whether it works.

precision A measure of the spread of quantitative results. If the measurements are precise all the results are very close in value.

prediction A statement about the expected outcome of a process. A prediction is not just a guess, it is based on scientific ideas.

pre-implantation genetic diagnosis (PGD) A type of gene technology. Embryos fertilised outside the body are tested for genetic variants that could cause disease. Only healthy embryos are put into the mother's uterus.

producer An organism at the start of a food chain. A producer is an autotroph, able to make its own food.

progesterone Hormone that is important in controlling the menstrual cycle.

prokaryotic organism An organism made of cells that do not contain a nucleus or organelles with membranes. Includes the simplest living organisms, such as bacteria.

protein A natural polymer made from amino acids. Proteins can be structural (e.g., collagen) or functional (e.g., enzymes).

protists Large group of eukaryotic microorganisms that can be single-celled or multicellular. Some protists cause disease.

Punnett square A genetic diagram that is used as a model of inheritance. It shows all the possible combinations of alleles of a gene, for the offspring of two individuals. It is used to make mathematical predictions about the possible genotypes of the offspring.

quadrat A square grid of a known area that is used when measuring the abundance of non-moving or slow-moving organisms in a location.

random error A measurement error due to results varying in an unpredictable way, for example, due to the scientist having to make a judgement about timing or colour.

range Describes the spread between the highest and the lowest of a set of measurements.

rate A measure of how quickly something changes. For example, the rate of a reaction can be investigated by measuring the formation of a product, such as a gas (in dm^3/s).

receptor A specialised cell that detects changes in an organism's environment. Also a specialised molecule on the surface of a cell that recognises and binds to a specific substance.

recessive You must have two copies of a recessive genetic variant to have the feature it produces. A recessive genetic variant will only show its associated feature when the genotype is homozygous.

rectum The final part of the digestive system before the anus.

red blood cell Type of blood cell containing haemoglobin, which transports oxygen around the body.

reflex arc A neuron pathway that brings about a reflex response. A reflex arc involves a sensory neuron, a relay neuron, and a motor neuron.

reflex response A rapid, involuntary response to a stimulus that uses nerve impulses transmitted along a reflex arc.

regulation The legal control of applications of science by a government or non-governmental body.

relay neuron A neuron that connects the sensory neuron to the motor neuron in a reflex arc.

repeatable Data are said to be repeatable when the same investigator finds the same or similar results under the same conditions. We can have more confidence in data that are repeatable.

representational model A type of scientific model that uses simple shapes or analogies to represent the interacting parts of a system. One example is the particle model – it helps us to visualise the tiny particles (atoms and molecules) that make up substances.

representative sample The characteristics of a representative sample are very similar to the characteristics of the whole population.

reproducible Data are said to be reproducible when other investigators have found the same or similar results under similar conditions. We can have more confidence in data that are reproducible.

resolution A microscope that lets you see details that are very close together has high resolution (resolving power).

response Action or behaviour that is caused by a stimulus.

risk An estimate of the probability that an unwanted outcome will happen. The size of the risk can be estimated from the chance of it occurring in a large sample over a given period of time.

risk factors Variables linked to an increased risk of disease (though they may not be the cause of the disease).

root hair cell Cell in a plant root that increases the surface area for absorption of water by osmosis and for minerals by active transport.

salivary gland Gland in the mouth that secretes digestive enzymes and saliva.

sample It is usually not possible to collect data about a whole population of organisms or other specimens. A study usually collects data about a proportion of them. This is a sample. Conclusions about a sample can only be applied to the whole population if it is a representative sample.

sanitation The hygienic disposal of waste, including faeces.

screening Testing large libraries of substances to find out which ones could be developed into new medicines. Screening is usually done by a machine.

secondary data Data collected by somebody else, which can be compared to the primary data collected in the lab or field by the person doing the investigation.

secretion The release of a hormone or other substance from a gland.

selective breeding The process of humans choosing organisms with certain characteristics and mating them to try to produce offspring with favourable characteristics.

sensory neuron Neuron that carries nerve impulses from a receptor to the CNS.

sex Whether an individual is male or female. Sex is determined by the inheritance of X and Y chromosomes. Sex should not be confused with gender.

sex chromosomes Chromosomes that determine an individual's sex. In humans it is chromosome pair 23 (the X and Y chromosomes).

sex hormone Hormones that control the development of male or female features. Androgens are one type of male sex hormone.

sexually transmitted infection (STI) A communicable disease caused by a pathogen, which can be caught and passed on during unprotected sex.

sickle-cell trait If a person is heterozygous for the recessive allele that causes sickle-cell disease, they have sickle-cell trait. This gives them some resistance to the protist pathogen that causes malaria.

significant figures The number of significant figures shows the precision that can be claimed for a piece of data. The first significant figure in a number is the first non-zero digit from the left.

sink In a plant, sugar is transported to a sink (e.g., a root or developing fruit) from a source (e.g., a leaf or storage organ).

small intestine Tube-like organ that receives food from the stomach, where digestion is completed and molecules from food are absorbed into the blood.

soil erosion Soil removal by wind or rain into rivers or the sea. This damages ecosystems. It is more likely after deforestation.

source (plants) In a plant, sugar is transported from a source (e.g., a leaf or storage organ) to a sink (e.g., a root or developing fruit).

spatial model A type of mathematical model in which a computer is used to make a model of one or more objects in a three-dimensional space. The model can be used to predict the outcome of changing a variable (e.g., temperature) in a given space (e.g., a landscape and the atmosphere above it).

specialised A cell that has differentiated to take on a specific job.

species A group of organisms that can breed to produce fertile offspring.

spores Resistant structures made by some bacteria and fungi that can survive in unfavourable conditions (e.g., cooking) and will grow again when conditions are more favourable.

stem cell Unspecialised animal cell that can divide and differentiate into specialised cells.

stem cell treatment Use of stem cells in medicine to treat damage or disease.

sterile Free from bacteria, fungi, protists, and viruses that could cause disease.

stimulus A change in an organism's environment that causes a response.

stomach Hollow organ that contains acid, and churns and breaks down food.

stomata (singular stoma) Tiny holes in the outer surface of a leaf that enable a plant to exchange gases (carbon dioxide, oxygen, and water vapour) with its surroundings.

subsistence farming Farming that produces only enough crop to support the farmer, the farmer's family, and the continuation of the farm.

substrate A substance that is changed by the action of a particular enzyme. Each molecule of the substrate must be the correct shape to fit into the enzyme's active site.

surface area The total area of a surface available for a chemical reaction or absorption to take place.

surface area:volume ratio The ratio of an organism's or object's surface area to its volume.

sustainability Using resources and the environment to meet the needs of people today without damaging Earth for people of the future. One way to do this is to use resources at the same rate as they can be replaced.

symptoms Feelings or changes experienced during illness (e.g., sore throat or runny nose).

synapse A tiny gap between neurons that transmits nerve impulses from one neuron to another using transmitter substances that diffuse across the gap.

system A collection of interacting parts.

systematic error A measurement error that differs from the true value by the same amount each time a measurement is made. A systematic error may be due to the environment, methods of observation, or the instrument used.

target An antigen or receptor molecule on a pathogen or host cell that could be affected by a medicine.

testosterone Hormone that is important in puberty in human males.

theory A scientific theory is a general explanation that applies to a large number of situations or examples (perhaps to all possible ones). It has been tested and used successfully, and is widely accepted by scientists. An example is the theory of evolution by natural selection.

thyroid gland Gland in the human body that secretes the hormone thyroxine.

thyroxine A hormone in the human body. It is secreted by the thyroid gland; it regulates growth and the rates of chemical reactions in cells.

transect A technique used to measure the distribution and abundance of organisms at regular intervals along a straight line through an ecosystem. Abiotic factors, including pollution levels, may also be measured.

transgenic An organism that has been genetically engineered to contain DNA from another organism.

transitional species A species that shows some features of an older species and some of a newer species. Fossils of transitional species show one possible way that the newer species could have evolved from the older one.

translocation The process of moving sugars, amino acids, and other substances through a plant.

transmitter substance Substance that diffuses across a synapse to transmit a nerve impulse from one neuron to another.

transpiration The process of water movement through a plant and its evaporation from cells in the leaves.

true value The actual value.

tumour A lump or ball of cells observed in patients with cancer. Changes (mutations) in genes that usually control the cell cycle cause a cell to divide many times by mitosis, forming a tumour.

turgid When a plant cell takes in water, it bulges and becomes stretched. It is said to be turgid. The strong cell wall prevents the cell from bursting.

type 1 diabetes Non-communicable disease in which the blood sugar level cannot be controlled effectively because pancreas cells do not make enough insulin.

type 2 diabetes Non-communicable disease in which the blood sugar level cannot be controlled effectively because pancreas cells do not make enough insulin, or the body's cells stop responding to it. Starts later in life, usually in people who are obese.

uncertainty An indication of the confidence a scientist has in the accuracy of a measurement. It can be expressed as a range of values within which the true value must lie.

unicellular An organism made of a single cell.

unspecialised A cell that has not yet differentiated to take on a specific job. Only stem cells are unspecialised.

urea Waste product of the breakdown of proteins and amino acids in the liver.

vaccination Introducing a substance (vaccine) into an organism to make it immune to a disease.

vaccine A vaccine contains dead or inactive pathogens, or parts of pathogens, used to trigger an immune response in an organism and establish immunity (without the organism catching the disease).

valid The conclusions from an experiment are valid if the procedures ensure that the effects observed are due to the cause claimed, and if the analysis has taken account of other possible factors.

valve Flap of tissue that acts like a one-way gate, only letting blood flow in one direction. Valves are found in the heart and veins.

variation Differences between the members of a group. For example, there is usually variation within a population of organisms and within repeated measurements of a quantity.

vector (biology) Used to transfer genes from one organism to another in genetic engineering. A plasmid, a bacterium, or a virus can be used as a vector.

vein Type of blood vessel that carries blood from cells and tissues to the heart.

ventricles The human heart is made of four muscular chambers. The ventricles are the lower chambers. The right ventricle pumps blood to the lungs and the left ventricle pumps blood to the rest of the body.

villi (singular villus) Finger-like projections on the wall of the human small intestine. Villi increase the surface area for absorption of substances from the gut.

viruses Large group of infectious agents that can only reproduce inside living cells. Some viruses cause disease.

water cycle Processes that cycle water through the abiotic and biotic parts of an ecosystem.

white blood cells Cells in the blood that are a major part of the immune system; one type of white blood cell ingests and digests pathogens; another type makes antibodies.

xylem Plant tissue that transports water and dissolved minerals through a plant.

Index

Acknowledgements

COVER: © The Trustees of the Natural History Museum, London

p2: Rawpixel/Shutterstock; **p4**: Saluha/iStockphoto; **p5(T)**: Aldo Murillo/iStockphoto; **p5(B)**: Dr Gopal Murti/Science Photo Library; **p6(T)**: Biophoto Associates/Science Photo Library; **p7**: Saluha/iStockphoto; **p8**: Spencer Grant/Science Photo Library; **p6(B)**: A Barrington Brown/Science Photo Library; **p10**: Wavebreak/iStockphoto; **p11(T)**: Sebastian Kaulitzki/Shutterstock; **p11(B)**: Kzww/Shutterstock; **p14(BR)**: Jani Bryson/iStockphoto; **p14(T)**: Andy Crawford/Getty Images; **p14(BL)**: PhotoAlto sas/Alamy; **p15(T)**: Dept of Clinical Cytogenetics, Addenbrookes Hospital/Science Photo Library; **p15(B)**: Edelmann/Science Photo Library; **p16(T)**: Ipatov/Shutterstock; **p17**: Mauro Fermariello/Science Photo Library; **p16(B)**: PeJo29/iStockphoto; **p18**: Squaredpixels/iStockphoto; **p20**: Alex-mit/iStockphoto; **p21(T)**: SnowWhite Images/Shutterstock; **p21(B)**: Tek Image/Science Photo Library; **p22**: Zmeel/iStockphoto; **p23(L)**: StockPhotosArt/iStockphoto; **p23(R)**: Photo courtesy of IRRI; **p25(T)**: Federico Rostagno/Shutterstock; **p25(B)**: Dr Neil Overy/Science Photo Library; **p26**: Tek Image/Science Photo Library; **p27**: Zmeel/iStockphoto; **p30**: Sebastian Duda/Shutterstock; **p32(T)**: Monkey Business Images/Shutterstock; **p32(B)**: Science Photo Library; **p36(T)**: Dr Kari Lounatmaa/Science Photo Library; **p36(TC)**: Eye of Science/Science Photo Library; **p36(BC)**: Eye of Science/Science Photo Library; **p36(B)**: London School of Hygiene & Tropical Medicine/Science Photo Library; **p37(T)**: Robynmac/iStockphoto; **p37(B)**: UK Crown Copyright - Courtesy of Fera; **p38(B)**: Schankz/Shutterstock; **p38(T)**: Monkey Business Images/Shutterstock; **p43(T)**: CDC/Science Photo Library; **p43(BR)**: Simply Signs/Alamy; **p44(T)**: Scott Camazine/Sue Trainor/Science Photo Library; **p44(B)**: Gyssels/Science Photo Library; **p45(T)**: UK Crown Copyright - Courtesy of Fera; **p45(BR)**: UK Crown Copyright - Courtesy of Fera; **p45(BL)**: Vlada Z/Shutterstock; **p47**: David Cole/Alamy; **p48**: Rawpixel Ltd/iStockphoto; **p49(T)**: Dina2001/iStockphoto; **p49(B)**: Ian Hooton/Science Photo Library; **p49(BL)**: Mkrberlin/Shutterstock; **p40**: Molekuul.be/Shutterstock; **p51(T)**: Monkey Business Images/Shutterstock; **p51(C)**: Science Photo Library; **p51(B)**: CNRI/Science Photo Library; **p52(T)**: Monkey Business Images/Shutterstock; **p52(B)**: Bilka/Shutterstock; **p53**: Christian Bertrand/Shutterstock; **p54**: RapidEye/iStockphoto; **p55(TL)**: Bettmann/Corbis; **p55(TR)**: Maxian/iStockphoto; **p55(B)**: Matt Meadows/Getty Images; **p57**: DepthofField/iStockphoto; **p58(T)**: CNRI/Science Photo Library; **p58(B)**: CNRI/Science Photo Library; **p59(T)**: Simon Fraser/Science Photo Library; **p59(B)**: Dr Stanley Flegler, Visuals Unlimited/Science Photo Library; **p60(T)**: Monkey Business Images/Shutterstock; **p60(B)**: Axel Bueckert/Shutterstock; **p62**: John Durham/Science Photo Library; **p63**: AJ Photo/Science Photo Library; **p64(T)**: Audrey Snider-Bell/Shutterstock; **p64(B)**: A and N photography/Shutterstock; **p65(T)**: Dlumen/iStockphoto; **p65(CL)**: Agencja Fotograficzna Caro/Alamy; **p65(CR)**: Matthew Chattle/Alamy; **p65(B)**: MichaelDeLeon/iStockphoto; **p68**: Dr Kari Lounatmaa/Science Photo Library; **p69**: Ian Hooton/Science Photo Library; **p72**: Lakov Kalinin/Shutterstock; **p74**: Jose Ignacio Soto/Shutterstock; **p75(T)**: Johan Swanepoel/Shutterstock; **p75(B)**: JC Revy, ISM/Science Photo Library; **p76**: Triff/Shutterstock; **p77(T)**: Oliver Hoffmann/Shutterstock; **p77(B)**: JC Revy, ISM/Science Photo Library; **p78(BR)**: British Antarctic Survey/Science Photo Library; **p78(BL)**: Derrick Neill/123RF; **p90(R)**: Lukasz Szwaj/Shutterstock; **p78(T)**: Gorkos/Shutterstock; **p90(L)**: F11photo/Shutterstock; **p83(L)**: Alistair Moore; **p83(C)**: Alistair Moore; **p83(R)**: Alistair Moore; **p86(TL)**: Biophoto Associates/Science Photo Library; **p86(TR)**: Biophoto Associates/Science Photo Library; **p86(B)**: Egilshay/Shutterstock; **p89(B)**: Biophoto Associates/Science Photo Library; **p89(TL)**: Kavring/Shutterstock; **p89(TR)**: VLDR/Shutterstock; **p90(R)**: Dr Jeremy Burgess/Science Photo Library; **p90(L)**: Madlen/Shutterstock; **p91(T)**: OZMedia/Shutterstock; **p91(B)**: Solarseven/Shutterstock; **p93**: Elena Elisseeva/Shutterstock; **p96**: Evoken/Shutterstock; **p98**: Robert Brook/Science Photo Library; **p100**: Martyn F Chillmaid/Science Photo Library; **p102(R)**: Luna Vandoorne/Shutterstock; **p102(L)**: U.S. Fish and Wildlife Service, Kirstin Breisch Russell; **p103**: Martin Shields/Science Photo Library; **p104**: Triff/Shutterstock; **p105**: Robert Brook/Science Photo Library; **p108**: Yuganov Konstantin/Shutterstock; **p102(T)**: Monkey Business Images/Shutterstock; **p102(B)**: Wavebreakmedia/Shutterstock; **p111**: KR Porter/Science Photo Library; **p112**: Peter Bernik/Shutterstock; **p113**: Janis Smits/123RF; **p114(T)**: Jason Lee/Shutterstock; **p114(C)**: Martyn F Chillmaid/Science Photo Library; **p114(B)**: Slawomir Fajer/Shutterstock; **p116**: Biophoto Associates/Science Photo Library; **p117(T)**: Mauro Fermariello/Science Photo Library; **p117(B)**: Dr David Furness, Keele University/Science Photo Library; **p118**: CC Studio/Science Photo Library; **p119**: Manfred Kage/Science Photo Library; **p121**: Eye of Science/Science Photo Library; **p122**: Neil Bromhall/Science Photo Library; **p124**: Nickolay Stanev/Shutterstock; **p125**: Ocskay Bence/Shutterstock; **p126(T)**: Mr Douzo/Shutterstock; **p126(C)**: Pascal Goetgheluck/Science Photo Library; **p126(B)**: Human Tissue Authority; **p128**: CC Studio/Science Photo Library; **p129(T)**: KR Porter/Science Photo Library; **p129(B)**: Mr Douzo/Shutterstock; **p132**: Malerapaso/iStockphoto; **p134**: Andrey_Popov/Shutterstock; **p139**: Iofoto/Shutterstock; **p141(T)**: Martin Dohrn/Royal College of Surgeons/ Science Photo Library; **p141(B)**: National Cancer Institute/Science Photo Library; **p142(R)**: Lebendkulturen.de/Shutterstock; **p142(L)**: Living Art Enterprises, LLC/Science Photo Library; **p143(R)**: Eye of Science/Science Photo Library; **p143(L)**: Steve Gschmeissner/Science Photo Library; **p145**: Maridav/Shutterstock; **p148(T)**: Iryna Tiumentseva/Shutterstock; **p148(B)**: Stuart G Porter/Shutterstock; **p149(T)**: Adam Hart-Davis/Science Photo Library; **p149(C)**: Adam Hart-Davis/Science Photo Library; **p149(B)**: Alexey Losevich/Shutterstock; **p150**: Henrik_L/iStockphoto; **p151**: Image Point Fr/Shutterstock; **p153**: Susan Schmitz/Shutterstock; **p154**: Stevecoleimages/iStockphoto; **p157**: Jean-Claude Revy-A Goujeon, ISM/Science Photo Library; **p159(T)**: Areeya_ann/Shutterstock; **p159(B)**: Mmkarabella/Shutterstock; **p160(A)**: Urbanbuzz/Shutterstock; **p160(B)**: Szefei/Shutterstock; **p160(C)**: Olga Popova/Shutterstock; **p160(D)**: Elena Elisseeva/Shutterstock; **p160(E)**: Mahathir Mohd Yasin/Shutterstock; **p160(F)**: 33333/Shutterstock; **p161(T)**: Samuel Ashfield/Science Photo Library; **p161(B)**: Mark Clarke/Science Photo Library; **p163**: Areeya_ann/Shutterstock; **p166**: Hanasch/Shutterstock; **p168(L)**: Pukhov Konstantin/Shutterstock; **p168(R)**: Marcel Kudláček/123RF; **p169(T)**: Erni/Shutterstock; **p169(C)**: Abi Warner/Shutterstock; **p169(B)**: Julianwphoto/iStockphoto; **p171(L)**: Acceptphoto/Shutterstock; **p171(C)**: Mark Winfrey/Shutterstock; **p171(R)**: Steve Heap/Shutterstock; **p172(L)**: Javier Brosch/Shutterstock; **p172(R)**: Holly Kuchera/Shutterstock; **p173(TC)**: Stephen Sharnoff/Visuals Unlimited, Inc.; **p173(TL)**: age fotostock/Alamy; **p173(TR)**: Joe Gough/Shutterstock; **p173(B)**: Paul D Stewart/Science Photo Library; **p175**: Chris Hellier/Science Photo Library; **p176(T)**: Dr Charlotte Brassey/Shutterstock; **p176(B)**: OUP Provided; **p177(B)**: © The Trustees of the Natural History Museum, London; **p179(TL)**: Tagstock1/Shutterstock; **p179(TR)**: Derrick Neill/123RF; **p179(B)**: Rudmer Zwerver/123RF; **p180**: Steve Gschmeissner/Science Photo Library; **p181(R)**: EdStock/iStockphoto; **p181(L)**: John Durham/Science Photo Library; **p182(TL)**: Georgios Kollidas/Shutterstock; **p182(TC)**: Christopher Swann/Science Photo Library; **p182(TR)**: Madlen/Shutterstock; **p182(B)**: Agustin Esmoris/Shutterstock; **p185**: International Union for Conservation of Nature (IUCN); **p186(T)**: Mary Ann McDonald/Shutterstock; **p186(BL)**: Tim Roberts Photography/Shutterstock; **p186(BR)**: Polygraphus/Shutterstock; **p187**: Nathape/Shutterstock; **p188(T)**: Markus Gann/Shutterstock; **p188(B)**: Borisoff/Shutterstock; **p189**: Chameleons Eye/Shutterstock; **p191(T)**: Huguette Roe/Shutterstock; **p191(B)**: Paul Rapson/Science Photo Library; **p192(TL)**: Hecke61/Shutterstock; **p192(TR)**: Adrian Arbib/Alamy; **p192(B)**: Cowboy54/Shutterstock; **p195**: Bildagentur Zoonar GmbH/Shutterstock; **p193(T)**: Robert Adrian Hillman/Shutterstock; **p193(TC)**: Black Sheep Media/Shutterstock; **p193(BC)**: Dutourdumonde Photography/Shutterstock; **p193(B)**: Massimo Brega, The Lighthouse/Science Photo Library; **p194**: Chris Hellier/Science Photo Library; **p198**: OUP Provided; **p200**: Alistair Moore; **p201(TL)**: Alistair Moore; **p201(TR)**: Alistair Moore; **p201(B)**: Alistair Moore; **p202(T)**: Alistair Moore; **p202(B)**: Alistair Moore; **p204(T)**: Steve McWilliam/Shutterstock; **p204(BC)**: Salpics32/Shutterstock; **p204(BL)**: Grigorii Pisotskii/Shutterstock; **p248(BR)**: Slhy/Shutterstock; **p206(T)**: Jiang Hongyan/Shutterstock; **p206(BL)**: OUP Provided; **p206(BR)**: OUP Provided; **p207(TL)**: OUP Provided; **p207(TC)**: OUP Provided; **p207(TR)**: OUP Provided; **p255(C)**: OUP Provided; **p255(B)**: OUP Provided; **p208(T)**: Image from the Central Science Laboratory Photographic Services Unit; **p208(B)**: Photo Africa/Shutterstock; **p209**: OUP Provided; **p210(TL)**: Alex-mit/iStockphoto; **p210(TR)**: Gyssels/Science Photo Library; **p210(CL)**: Martin Shields/Science Photo Library; **p210(CR)**: Jason Lee/Shutterstock; **p210(BL)**: Zephyr/Science Photo Library; **p210(BR)**: Pal Teravagimov/Shutterstock; **p211**: Martyn F Chillmaid/Science Photo Library; **p214**: Chris Hellier/Science Photo Library; **p217**: Agencja

Fotograficzna Caro/Alamy; **p218**: Volt Collection/Shutterstock; **p223**: Trevor Clifford Photography/Science Photo Library; **p224**: Martyn F Chillmaid/Science Photo Library; **p245(B)**: Martyn F Chillmaid/Science Photo Library; **p246(T)**: Silvano Audisio/Shutterstock; **p247(T)**: Vladimir Prusakov/Shutterstock; **p247(B)**: Imagedb.com/Shutterstock; **p246(C)**: Thinglass/Shutterstock; **p246(B)**: AndrewJShearer/iStockphoto; **p248(T)**: Photolinc/Shutterstock; **p248(B)**: Edward Shaw/iStockphoto; **p245(T)**: Francesco Abrignani/Shutterstock;

OXFORD
UNIVERSITY PRESS

Great Clarendon Street, Oxford, OX2 6DP, United Kingdom

Oxford University Press is a department of the University of Oxford. It furthers the University's objective of excellence in research, scholarship, and education by publishing worldwide. Oxford is a registered trade mark of Oxford University Press in the UK and in certain other countries

Resources developed by the University of York Science Education Group on behalf of Salters' Educational Resources Ltd

The moral rights of the authors have been asserted

First published in 2016

British Library Cataloguing in Publication Data
Data available

978 0 19 835950 0

10 9 8 7 6 5 4 3 2 1

Paper used in the production of this book is a natural, recyclable product made from wood grown in sustainable forests. The manufacturing process conforms to the environmental regulations of the country of origin.

Printed in Great Britain by Bell and Bain Ltd. Glasgow

This resource is endorsed by OCR for use with specification J260 OCR GCSE (9–1) in Combined Science B (Twenty First Century Science). In order to gain OCR endorsement, this resource has undergone an independent quality check. Any references to assessment and/ or assessment preparation are the publisher's interpretation of the specification requirements and are not endorsed by OCR. OCR recommends that a range of teaching and learning resources are used in preparing learners for assessment. OCR has not paid for the production of this resource, nor does OCR receive any royalties from its sale. For more information about the endorsement process, please visit the OCR website, www.ocr.org.uk.

Project Team acknowledgements

These resources have been developed to support teachers and students undertaking the OCR suite of specifications GCSE Twenty First Century Science. They have been developed from the first and second editions of the resources.

We would like to thank Michelle Spiller, Sarah Old, Naomi Rowe, and Ann Wolstenholme at OCR for their work on the specifications for the Twenty First Century Science course.

We would also like to thank the following contributors to case studies in these resources: Judith Hogarth and Susan Wood (Royal Victoria Infirmary); Sarah-Louise Hayes (Freeman Hospital); Charlotte Brassey and Sally Collins (Natural History Museum); Adrian Fox, Charles Lane, and Julian Smith (Fera Science Ltd).

Authors and editors from the first and second editions

We thank the authors and editors of the first and second editions: Jenifer Burden, Cris Edgell, Ann Fullick, Anna Grayson, Angela Hall, Bill Indge, Neil Ingram, Mike Kalvis, Pam Large, Carol Levick, Jean Martin, Nick Owens, Cliff Porter, Jacqueline Punter, Anne Scott, Carol Usher, and Linn Winspear.

Many people from schools, colleges, universities, industry, and the professions contributed to the production of the first edition of these resources. We also acknowledge the invaluable contribution of the teachers and students in the pilot centres.

A full list of contributors can be found on the website: https:// global.oup.com/education/content/secondary/online-products/ acknowledgements/?region=uk

The first edition of Twenty First Century Science was developed with support from the Nuffield Foundation, The Salters' Institute, and the Wellcome Trust.

The continued development of Twenty First Century Science is made possible by generous support from The Salters' Institute.